Commercial Heating
First Edition – Revised

CW00394391

Technical Author: Colin Poole

Technical Editor: Chris Long

Technical Illustrator: Chris Long

Previous contributors:
          CORGI Technical Team
          GasForce Limited
          British Gas Services Ltd
          Ambi-Rad Ltd
          Andrews Water Heaters
          Monodraught Ltd

CORGI*direct* has used its best efforts to prepare this manual, but makes no warranty about the content of the manual and will not be liable under any circumstances for any direct or indirect damages resulting from any use of this manual.

Acknowledgement

Our special thanks goes to Colin Poole whose significant contribution and support during the original drafting of this manual have greatly assisted CORGI*direct*.

CORGI*direct* gratefully acknowledges the use in this manual of reference material published by the British Standards Institution (BSi) and the Institution of Gas Engineers and Managers (IGEM).

Published by

CORGI*direct*

CORGI*direct*

Telephone: 0800 915 0490
Website: www.corgi-direct.com

First Edition: September 2007

Reprinted: August 2012

Printed by: Blackmore Ltd

ISBN: 978-1-907723-12-4

Contents

Contents

Contents

Contents

13 – **CORGI***direct* **Publications**

General Introduction

The Commercial Heating – Non-Domestic (First Edition – Revised) manual builds on the information contained within Essential Gas Safety – Non-Domestic, from the Gas Installer Manual – Non-Domestic Series.

Generic information covering combustion, ventilation, flueing, installation of pipework and fittings, gas tightness testing and purging, etc. are purposefully left out of this Commercial Heating – Non-Domestic manual as these subjects are covered in greater detail within Essential Gas Safety – Non-Domestic.

However, specific information regarding appliance types and their uses, installation including flueing and ventilation requirements, generic commissioning, servicing and fault finding for a given heating appliance are discussed within this manual.

This revised First Edition takes account of changes to primary legislation since the manual was first printed and where appropriate, reviews and updates information contained within each part. However, the appliance fundamentals – how they work, their general installation requirements, etc. has not changed significantly and therefore remains applicable to todays gas installer.

It is important to note that this manual does not cover all the requirements of legislation and the multiple industry standards involved and should therefore not be read in isolation or regarded as an alternative to the source documents.

Heating/hot water design requirements for commercial premises – 1

Introduction

Gas, and in particular Natural gas (NG), has over the past three decades become increasingly popular as a primary fuel for space heating in non-domestic buildings. What was once a market dominated by large centralised heating systems, usually fuelled by coal or oil, has developed to that of mainly de-centralised individually heated areas where gas can be used for all the premium advantages that it has to offer, namely:

*   sulphur free

*   clean burning

*   reduced maintenance

*   reduced manual handling

*   delivered to point of use, etc.

However, the development of the gas market is not without long term cost. After all, gas has a finite availability and it is important that this resource is used to its best advantage whilst it is available for the benefit of this and future generations.

Coincidental to this development in the gas utilisation market, is the world wide increase in carbon emissions from all forms of fossil fuel usage that led to an international treaty for reduction of carbon emissions in to the atmosphere.

UK government directives have evolved, working towards the overall reduction of carbon emissions. Underpinning these requirements is the need for all involved with fossil fuel usage, whether architects, builders, plumbers, gas installers or building owners, to have a general awareness of fossil fuel efficient heating systems.

This manual seeks to review the principle methods of gas heating available for non-domestic premises and to discuss the relative merits of one appliance type or system against others.

In parallel with this objective is the requirement that all gas installers are competent in the type of activity they carry out in their day-to-day employment. This manual also seeks to guide operatives in the safe installation and maintenance of the various heating systems available.

The information contained in this manual has been developed after gathering data and information from various sources, which CORGI*direct* believes reflects current custom and practice within the non-domestic heating market.

General requirements

The appliances reviewed in the various sections will need to comply with legislation and British/European Standards applicable at their date of manufacture and time of installation, including the Gas Appliance (Safety) Regulations (GASR).

Appliances will normally have an easily legible data plate, bearing information that indicates the type of gas to be used, the gas inlet pressure and where necessary, gas regulator pressure setting for the appliance:

*   NG (Natural Gas) – 20mbar inlet and for example, 12mbar burner pressure

*   P (Propane) – 37mbar inlet, 25mbar burner pressure

*   B (Butane) – 28mbar inlet, 21mbar burner pressure.

Gas operatives will need to ensure that the appliance meets the manufacturer's original specifications; if modifications have been made and if those modifications have been undertaken using the appropriate parts.

In relation to second hand (i.e. previously used) appliances, the information contained in this manual can also be applied, providing the appliance is serviceable and safe to use.

It is a requirement of the Gas Safety (Installation and Use) Regulations (GSIUR) that no person shall carry out 'work' in relation to gas appliances and other gas fittings unless they are 'competent' to do so.

It is also a requirement of the GSIUR that any gas businesses who carry out 'work' on gas appliances and other gas fittings in domestic and commercial premises are registered with a body approved by the Health and Safety Executive (HSE), namely Gas Safe Register.

The valid certificates of competence for individual gas fitting operatives will need to have been issued under the Nationally Accredited Certification Scheme (ACS) for the area of activity in which they operate (see **Part 3 – Competency** for further guidance).

Note: No employer, member of the public or other responsible person should knowingly employ a person who works on gas and yet does not comply with the aforementioned requirements.

All non-domestic gas appliances and other gas fittings need to be installed in accordance with the GSIUR, relevant Building Regulations and manufacturers' instructions, except where specific exceptions apply. Installations will also need to comply with the requirements of British Standards together with other industry standards (as appropriate) and regulations for electrical installations or those regulations appropriate to the geographical region in which they are installed.

In all instances where an appliance is supplied and fitted, a copy of the manufacturer's installation, user instructions and any other information provided with the appliance needs to be left with the responsible person and/or end user on completion of the installation works.

This requirement is a stipulation of Regulation 29 of the GSIUR.

Types of heating systems

Hot water heating systems

Water boilers and their various systems are one of the most common forms of gas utilisation for heating purposes in domestic, commercial and industrial properties.

There is a wide variation in size of hot water central heating systems that the non-domestic gas operative can encounter, from small public houses up to large commercial and industrial sites.

They are frequently used throughout all market sectors in the generation of hot water for space heating purposes (central heating) and additionally, in the commercial sectors for bulk hot water uses (see **Part 4 – Hot water heating systems** for further guidance).

Industrially, boilers are also used in the production of steam for a variety of processes and power generation. Whilst steam raising boilers are usually beyond the scope of the ACS, in terms of operative competency requirements, this manual offers an introduction to the types of steam raising boilers that can be commonly found on some industrial sites other than for power generation (see **Part 6 – Boilers** for further guidance).

Overhead radiant heating

Overhead radiant heating can offer significant energy savings compared with some other forms of heating systems such as centralised boiler systems, or even some warm air distribution systems. This is due to the design ability to target heat where it is needed, particularly in large working areas or where there is a high level of air volume changes per hour (ac/hr).

Radiant heating can be used as an effective method of spot heating, for example in a large warehouse area where staff operating in the warehouse are localised in one particular area within the warehouse.

Alternatively, by building a series of heater units to cover the whole floor area, it may be used as a total heating scheme.

An area may be heated for the general comfort of the occupants whilst maintaining a cooler ambient temperature than would otherwise be maintained using alternative systems of heating, which heat the total volume of the building, including the roof space which is not occupied (e.g. warm air distribution).

Other benefits include:

- rapid heat-up times

- no floor space is occupied by the heating units; and

- with multiple burner systems particularly, the loss of a single burner under fault conditions would not usually cause significant loss of room temperature (see **Part 7 – Overhead radiant heating** for further guidance).

Warm air heating
– indirect and direct-fired

Indirect and direct-fired gas heaters provide hot air into a room or workspace by the principle of convection. Convective heat transfer relies on the movement of a gas (air) or fluid to carry heat with it and then transfers the heat to other gases, fluids or solids.

Warm air heating systems tend to be universally acceptable to most occupants and therefore tend to be commonly used for speculative warehouse or factory type developments.

With correct design, using the most appropriate systems for the application (including the use of de-stratification fans) significant energy savings may be made, compared to traditional centralised wet boiler systems.

The systems available are very versatile and offer other tangible benefits such as floor space saving with suspended systems, good accessibility with minimum maintenance downtime and rapid heat-up times (see **Part 8 – Warm air heating systems** for further guidance).

Combined heat and power (CHP)

CHP systems represent proven technology whereby a gas-fuelled engine is used as the motive force to create power.

By using numerous heat exchangers in the system, it is possible to recover most of the wasted energy from the engine jacket and from the exhaust gases to provide heating and hot water for a building. In this way, the energy used in the power generation process can be fully utilised to provide combined heat and power (CHP).

CHP systems are capable of achieving very high system energy efficiencies and have come to the fore in recent years as a result of global initiatives to lower carbon emissions in the atmosphere.

It is essential for any CHP system to be correctly designed to ensure continued use of all energy and heat produced by the unit. They will therefore usually be designed to satisfy the constant base load for power, heating and hot water at any given site with conventional systems used to 'top up' and match any load requirement in excess of the base demand (see **Part 9 – Combined heat and power** for further guidance).

Gas and associated legislation – 2

2 – Gas and associated legislation

Introduction

When installing, servicing, maintaining or repairing gas appliances and other gas fittings in Great Britain, Northern Ireland and the Isle of Man, there are a number of statutory requirements with which to comply. They cover all aspects of work including but not limited to; gas, electricity, water and building work. When working on gas, compliance with legislation should be achieved by carrying out work in accordance with the relevant industry standards and manufacturers' installation instructions.

When any work (see **Part 11 – Definitions**) is carried out in relation to gas appliances and other gas fittings covered by this manual, gas operatives must be competent and hold a valid certificate of competence for each work activity that they wish to undertake. The valid certificate must have been issued under the National Accredited Certification Scheme (ACS) for individual gas fitting operatives, or through a Gas Services National/Scottish Vocational Qualification (N/SVQ), managed by Energy & Utility Skills (see **Part 12 – References**).

All gas appliances and other gas fittings must be installed in accordance with the Gas Safety (Installation and Use) Regulations, Building Regulations and manufacturer's installation instructions.

Installations should also comply with the requirements of British/European Standards together with other industry standards (as appropriate) and regulations for electrical installations or those regulations appropriate to the geographical region in which they are to be installed.

General

The Gas Safety (Installation and Use) Regulations 1998 (England, Scotland & Wales); The Gas Safety (Installation and Use) Regulations (Northern Ireland) 2004; Gas Safety (Installation and Use) Regulations 1994 as amended & applied by the Gas Safety (Application) Order 1996 (as applied to Isle of Man) is the primary legislative reference document for all businesses and their operatives involved with the safe installation, maintenance and use of gas systems. Including gas fittings, appliances and flues, mainly in domestic and commercial premises, e.g. offices, shops, public buildings and similar places.

Note: For detailed guidance regarding the principle regulations for England, Scotland & Wales (the GSIUR) in establishments, see Approved Code of Practice and Guidance – Safety in the installation and use of gas systems and appliances – Gas Safety (Installation and Use) Regulations 1998, (Order Ref: GR1, see Part 13 – CORGI*direct* Publications).

When work is to be carried out on non-domestic premises there are also other important items of legislation that interface with the GSIUR, which will also need to be complied with, such as (the list isn't exhaustive):

• Health and Safety at Work etc. Act 1974 (HSWA) United Kingdom

• Health and Safety at Work (Northern Ireland) Order 1978

• Management of Health and Safety at Work Regulations 1999 (MHSWR)

• Gas Acts 1986 and 1995 (GA)

• Pipelines Safety Regulations 1996 (PSR)

• Pressure Equipment Regulations 1999 (PER) United Kingdom

- Pressure Equipment Regulations (Amendment) Regulations 2002 (United Kingdom)

- Gas Safety (Management) Regulations 1996 (GSMR)

- Workplace (Health, Safety and Welfare) Regulations 1992 (WHSWR)

- Provision and Use of Work Equipment Regulations 1998 (PUWER)

- Gas Appliances (Safety) Regulations 1995 (GASR)

- Construction (Design and Management) Regulations 2007 (CDM) Great Britain and Northern Ireland

- Construction (Design and Management) Regulations 2003 (as applied to Isle of Man)

- Pressure Systems Safety Regulations 2000 (PSSR) Great Britain

- Pressure Equipment (Amendment) Regulations 2002 (United Kingdom)

- Health and Safety (Safety Signs and Signals) Regulations 1996 (HSSSR)

- Reporting of Injuries, Diseases and Dangerous Occurrences Regulations 1995 (RIDDOR) Great Britain

- RIDDOR 1997 (Northern Ireland)

- RIDDOR 1999 (Isle of Man)

- The Dangerous Substances & Explosive Atmospheres Regulations 2002 (DSEAR).

Note: For general information regarding these regulations, see Essential Gas Safety – Non-Domestic (order ref: ND1, see Part 13 – CORGI*direct* Publications).

This manual will review some of the particular requirements of other legislation that have an impact on non-domestic properties, namely:

- The Factories Act

- Control of Substances Hazardous to Health Regulations 2002 (COSHH)

- Building Regulations for England and Wales

- Building requirements for other areas of the UK.

The Factories Act

The GSIUR applies to any building defined as premises in section 53 of the HSWA 1974. However, with the exception of Regulation's 37 (Escape of gas), 38 (Use of antifluctuators and valves), 41 (Revocation and amendments) and those areas used as sleeping accommodation, the GSIUR generally excludes factories within the meaning of the Factories Act 1961.

The expression 'factory' is defined within the Factories Act 1961, 175 as –

'any premises in which, or within the close or curtilage or precincts of which, persons are employed in manual labour in any process for or incidental to any of the following purposes, namely:

a) the making of any article or part of any article; or

b) the altering, repairing, ornamenting, finishing, cleaning or washing or the breaking up or demolition of any article; or

c) the adapting for sale of any article; or

d) the slaughtering of cattle, sheep, swine, goats, horses, asses or mules; or

e) the confinement of such animals as aforesaid while awaiting slaughter at other premises...'

In addition to factory buildings defined above, sections 123-126 of the Factories Act 1961, classifies certain other places such as docks and electrical power stations as factories.

Where any particular premises such as factories are specifically excluded from the GSIUR there will be other legislative provisions such as the HSWA 1974, which do apply and require similar gas safety requirements.

The Factories Act 1961 is allied to the HSWA 1974, in that it sets out the requirements for the general safety of personnel working within the confines of a factory premises.

Some of the subject areas covered by the Factories Act 1961 that can be associated with working on gas installations include:

*   dangerous substances

*   dangerous fumes and lack of oxygen

*   precautions with respect to explosive or flammable dust, gas vapour or substance

*   steam boilers

    -   attachments and construction

    -   maintenance, examination and use

    -   restrictions on entry

*   precautions as respects water sealed gas holders.

In order to satisfy the requirements of the Factories Act and the HSWA it should be recognised that the GSIUR provides the basis for the safe working on gas systems, even at premises such as factories that are specifically excluded.

With this in mind the Health and Safety Executive (HSE) has approved and published an 'Approved Code of Practice' (ACOP) guidance to the GSIUR, which is available from HSE Books (see **Part 12 – References**) or alternatively through CORGI*direct*, quoting order reference GR1, see **Part 13 – CORGI*direct* Publications**.

Control of Substances Hazardous to Health Regulations 2002 (COSHH)

The following information on COSHH draws upon freely available information produced by the HSE such as 'Working with substances hazardous to health. What you need to know about COSHH' – coded: INDG136(rev4).

Copies of this and other information sheets can be obtained from the HSE's dedicated knowledge resource – www.hse.gov.uk/coshh.

Using chemicals or other hazardous substances at work, as well as engaging in certain work activities can put people's health at risk. Therefore, the law requires that employers control (risk assess to remove or control the hazard) exposure to hazardous substances to prevent ill health.

Employers, including self-employed have to protect both employees and others who may be exposed, by complying with the Control of Substances Hazardous to Health Regulations 2002 (COSHH) (as amended).

COSHH defines hazardous substances to include:

*   substances used directly in work activities, e.g. adhesives, paints, cleaning agents, etc

*   substances generated during work activities, e.g. fumes from soldering and welding

*   naturally occurring substances, e.g. grain dust

*   biological agents such as bacteria and other micro-organisms.

COSHH applies to virtually all substances hazardous to health except:

- asbestos and lead which have their own regulations
- substances which are hazardous only because they are;
  - radioactive; or
  - at high pressure; or
  - at extreme temperatures; or
  - have explosive or flammable properties (other regulations apply to these risks)
- biological agents that are outside the employer's control, e.g. catching an infection from a workmate.

For the vast majority of commercial chemicals, the presence (or not) of a warning label will indicate whether COSHH is relevant, e.g. there is no warning label on ordinary household washing-up liquid, so there is no need to worry about COSHH. However, there is a warning label on bleach and so COSHH does apply to its use in the workplace.

As a result of this legislation, when a gas operative is preparing to visit a non-domestic site, COSHH statements should be available for all relevant substances the operative needs, in order to carry out their work. This is because the COSHH statements may be requested by the clerk of works or health and safety representative for the site as part of the site access permit regime.

Note: Further information on COSHH and its requirements can be obtained from the HSE website and via HSE Books (see Part 12 – References).

Building Regulations (England and Wales) 2010

Building Regulations exist to ensure the health and safety of people occupying domestic, commercial and industrial premises.

The regulations contain various sections dealing with definitions, procedures and what is expected in terms of the technical performance of building work, e.g. they define what types of building, plumbing and heating projects are defined as 'building work' and are therefore subject to control under the Regulations. They also provide regulation for energy conservation of the buildings.

The primary responsibility for complying with the regulations rests with the person carrying out the work, however it is ultimately the building owner who would be served with any enforcement notice if the work does not comply with the regulations.

Practical guidance on ways to comply with the functional requirements of the Building Regulations is contained in a series of 'Approved Documents', which are to be read in conjunction with each of the fourteen parts of the regulations (see Table 2.1).

Gas operatives will therefore need to be familiar, particularly with part 'J ' and part 'L', but also when carrying out certain electrical works, part 'P'.

The efficiency of heating systems fitted in buildings used for commercial and industrial purposes are covered by Approved Documents:

- L2A (ADL2A) Conservation of fuel and power (New buildings other than dwellings); and
- L2B (ADL2B) Conservation of fuel and power (Existing buildings other than dwellings).

These two documents give guidance on how to satisfy the energy performance requirements in non-domestic buildings.

Table 2.1 The fourteen parts of schedule 1 to the building regulations

Part	Description
A	Structure
B	Fire Safety
C	Site preparation and resistance to contaminants and moisture
D	Toxic substances
E	Resistance to the passage of sound
F	Ventilation
G	Sanitation, hot water safety and water efficiency
H	Drainage and waste disposal
J	Combustion appliances and fuel storage systems
K	Protection from falling, collision and impact
L	Conservation of fuel and power
M	Access to and use of buildings
N	Glazing – safety in relation to impact, opening and cleaning
P	Electrical safety

The 'Non-Domestic Building Services Compliance Guide' is a second tier document referred to in ADL2A and ADL2B as a source of guidance for complying with the requirements of the Building Regulations Part L for space heating systems, hot water systems, cooling and ventilation systems. The guide outlines minimum provisions for compliance with Part L for each type of heating, hot water, cooling or air distribution systems.

Seasonal efficiency and SEDBUK

Heat generator seasonal efficiency is the estimated seasonal ratio of heat input to heat output from the heat generator. The seasonal efficiency will depend on the operating mode of the heat generator over the heating season.

For boiler systems it is based on a weighted average of the efficiencies of the boiler at 15%, 30% and 100% of the boiler output.

For the calculation method to determine seasonal efficiencies of commercial heating systems, gas operatives will need to refer to the Non-Domestic Building Services Compliance Guide and probably competent third parties, i.e. registered energy assessors (see **Standard Assessment Procedure** in this Part).

Seasonal Efficiency for Domestic Boilers in the UK (SEDBUK) was developed under the Government's Energy Efficiency Best Practice Programme with the co-operation of domestic boiler manufacturers and provides a basis for fair comparison of the energy performance of different boilers.

SEDBUK is the average annual efficiency achieved in typical domestic conditions, making reasonable assumptions about pattern of usage, climate, control and other influences.

Figure 2.1 SEDBUK 2005 Bands

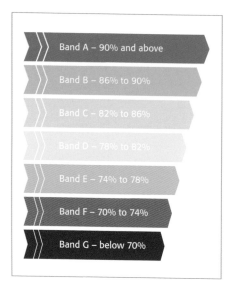

Band A – 90% and above

Band B – 86% to 90%

Band C – 82% to 86%

Band D – 78% to 82%

Band E – 74% to 78%

Band F – 70% to 74%

Band G – below 70%

It is calculated from the results of standard laboratory tests together with other important factors such as boiler type, ignition arrangement, internal store size, fuel used, knowledge of the UK climate and typical domestic usage patterns.

The purpose of the project was to develop a method for calculating seasonal efficiency for individual central heating boilers to use in the UK Building Regulations – Part L.

As a simple guide to efficiency, a scheme was created with SEDBUK efficiency bands, assigned to boilers on an 'A' to 'G' scale (see Figure 2.1 SEDBUK 2005 bands).

Since its introduction in 1999, SEDBUK has been through a number of changes – recently SEDBUK 2005 to the latest version, SEDBUK 2009.

Correspondingly, SAP 2005 (see **Standard Assessment Procedure** in this Part) uses data from SEDBUK 2005 and SAP 2009 use that of SEDBUK 2009.

The old 'band' is shown in the database and may be used on product literature and labels, though there is no requirement for manufacturers to do so.

The SEDBUK bands were withdrawn on 1st October 2010 so as not to cause confusion with the proposed Energy using Products Directive (EuP), which will introduce appliance labelling in the same vein as that currently used for 'white goods'.

SEDBUK 2009 uses only the percentage figure and its methodology has been tightened, which introduces a maximum ceiling of 88% efficiency.

Standard Assessment Procedure (SAP)

SAP is the process used for assessing the energy performance of dwellings based on the energy costs associated with space heating, water heating, ventilation and lighting, less cost savings from energy generation technologies.

Note: Where part of an accommodation unit is used for commercial purposes, e.g. an office or shop, this part of the building should be included as part of the dwelling if it could revert to domestic use on a change of occupancy.

Where the commercial part of the building would not be expected to revert to domestic use, the dwelling is to be assessed by SAP and the remainder by procedures for non-domestic buildings.

The indicators of energy performance for a dwelling are:

* energy consumption per floor area

* energy cost rating (SAP rating)

* environmental impact
  (based on CO_2 emissions)

* dwelling CO_2 emission rate (DER).

The SAP rating of a dwelling is expressed on a scale up to 100, where the higher the SAP rating number, the lower the running costs.

The SAP calculation is based on a range of factors (not affected by geographical location) that contribute towards the energy efficiency of the dwelling, such as:

* construction material and its thermal insulation value

* ventilation system

* efficiency of the heating system and its method of control

* solar gain

* type of fuel used for space heating, water heating, ventilation and lighting

* renewable energy technologies used.

The dwelling energy rating is based on the annual CO_2 emissions per unit floor area for space heating, water heating, ventilation and lighting, less cost savings from energy generation technologies and is expressed as a figure in $kg/m^2/year$.

For wholly non-domestic buildings (non-dwellings), compliance with Building Regulations in terms of energy efficiency is demonstrated/measured against the Energy Performance of Buildings Directive (EPBD), which calculates the annual energy usage of the proposed building and compares it with a 'notional' building.

This is achieved using a simplified tool developed by BRE (Building Research Establishment) on behalf of the DCLG.

The tool entitled 'Simplified Building Energy Model' (SBEM) is a computer program that provides analysis of a building's energy consumption. This then leads to the generation of Energy Performance Certificates (EPC) & Display Energy Certificates (DEC) for public buildings.

As mentioned previously, EPC/DEC and the use of SBEM is the preserve of registered energy assessors/consultants and not typically gas operatives/businesses. However, the Non-Domestic Building Services Compliance Guide is a useful tool and should therefore be referred to by gas businesses/operatives.

Energy saving control systems

The Non-Domestic Building Services Compliance Guide defines the minimum controls package, specific to each technology, to reduce carbon emissions.

Any additional controls that may be installed on any particular system that represents more than the minimum requirements can be given a measure of efficiency credit when calculating the Effective Heat Generating Seasonal Efficiency.

Controls referred to in the aforementioned guide include:

* **timing and temperature control** – a means of controlling the start and stop time for the system, in conjunction with zone specific temperature control during 'on' periods

* **black bulb thermostat** – a means of temperature control associated with areas heated by radiant systems, which will allow comfort conditions to be maintained at lower air temperatures

* **high limit thermostat** – a control mechanism to restrict the temperature of an appliance or system to prevent possible appliance damage or hazard as a result of overheating

* **weather compensation** – a system of control, which assesses outside temperature, inside temperature and system hot water return temperature (on boiler systems). The return hot water is re-circulated independent of the boiler until, the system hot water return temperature drops below a preset limit to bring the boilers on

- **optimal start/stop** – a system of control, which assesses outside temperature, inside temperature and retained knowledge of the system efficiency. The boiler start is delayed to the latest required in achieving set room temperature by preset occupancy time. Similarly, the controller will shut down the boiler at a time it determines will be the earliest to maintain room temperature until the preset latest time of occupancy

- **night set back (frost protection)** – a control thermostat to allow the system to operate at lower temperatures during periods of no occupancy to that required as a minimum for protection against possible effects of extreme low temperature, e.g. frost damage

- **two stage firing** – a control mechanism that will permit the system to operate at a high and low firing capacity as necessary to match the variable demands during periods of occupancy. This can be either as high/low firing on an individual appliance or as a function of multiple appliances, e.g. modular boilers

- **heat loss limitation (modular boilers)** – a control mechanism to reduce heat loss from non-firing boilers such as automatic damper mechanism in the flue exit from the appliance, which closes when that particular boiler is not 'on line'

- **sequence control (modular boilers)** – a control mechanism to sequence boiler start/stop based on load requirements at any particular time

- **modulating control** – a control mechanism to permit the burner of a single appliance to modulate to meet the load requirements at any particular time

- **economiser** – a device included in a secondary heat exchanger fitted on a boiler system providing additional heat transfer capacity.

Boilers in new and existing buildings

It is a requirement of the Part L that any new building is provided with high efficiency (HE) boilers (typically these will be condensing appliances, but certain non-domestic condensing appliances may also be appropriate) to meet the minimum requirements of ADL2A and ADL2B.

Commercial buildings are recognised as being more difficult to heat using condensing boiler methods compared to domestic buildings and therefore due consideration is placed on the physical restraints that may apply in individual circumstances. This is reflected in the calculation for Effective Heat Generating Seasonal Efficiency (EHGSE).

Where a single or multiple NG boilers are used to meet the heat demand for a new building, the boiler(s) will need to achieve a boiler seasonal efficiency of not less than 86% of gross calorific value (CV).

For existing buildings, the minimum requirement is 84% of gross CV. Each boiler will need to achieve a boiler seasonal efficiency of not less than 82% of gross CV and the overall effective boiler seasonal efficiency of not less than 84% of gross CV.

If the seasonal efficiency of a replacement boiler does not meet the minimum required EHGSE for that type of boiler, additional measures will need to be adopted by improving the control features of the system.

As previously mentioned in this Part, under **Energy saving control systems** – any additional controls installed above the minimum requirements can be given a measure of efficiency credit when calculating the EHGSE.

Some examples of heating efficiency credits for boilers are as follows:

- multiple boilers   1%
- sequential control of multiple boilers   1%
- thermostatic radiator valves   1%
- optimised start and stop   2%
- weather compensation   2%

For example, a boiler of rated seasonal efficiency of 80% fitted with optimised start and stop and weather compensation would accrue an additional 4% heating efficiency credit, which would then bring it up to the minimum required of 84%.

Warm air heaters in new and existing buildings

The Non-Domestic Building Services Compliance Guide defines the types of gas-fired warm air systems into 3 basic types:

- type 1 – forced convection, indirect-fired natural draught burner
- type 2 – forced convection, indirect-fired forced or induced draught burner
- type 3 – forced convection, direct-fired burner.

Unlike boilers, warm air heaters operate under the same conditions at all times so the EHGSE is deemed to be the same as their steady state thermal efficiency based on gross CV as declared by the manufacturers.

In order to meet the requirements of ADL2A and ADL2B, warm air systems in new and existing buildings will need to achieve EHGSE of 91% (net CV) for types 1 and 2 appliances and 100% (net CV) for type 3 appliances.

In any case, the minimum controls package will need to include time and temperature control; in addition, any building greater than 150m^2 in total floor area will need to include zone control.

As previously stated, any additional controls installed above the minimum requirements can be given a measure of efficiency credit when calculating the EHGSE.

Some examples of heating efficiency credits for warm air heaters are as follows:

- optimised stop   1%
- two stage burners (high/low)   2%
- modulating burners   3%

For example a type 1 heater of rated seasonal efficiency of 87% (net CV) fitted with optimised stop and modulating burner would accrue an additional 4% heating efficiency credit, equating to the minimum 91% required.

Overhead radiant heaters in new and existing buildings

The Non-Domestic Building Services Compliance Guide defines the types of gas-fired radiant systems into 3 basic types:

- luminous
- non-luminous
- multi-burner.

Like warm air heaters, radiant heaters operate under the same conditions at all times so the EHGSE is deemed to be the same as their steady state thermal efficiency based on gross CV as declared by the manufacturers.

In order to meet the requirements of ADL2A and ADL2B radiant systems in new and existing buildings will need to achieve EHGSE of 86% for unflued luminous and non-luminous appliances, 86% for flued non-luminous appliances and 91% for multi-burner systems.

In any case, the minimum controls package will need to include time and temperature controls with black bulb sensors.

If the seasonal efficiency of a radiant heater does not meet the minimum required EHGSE for that type of heater, additional measures will need to be adopted by improving the control features of the system. As previously stated, any additional controls installed above the minimum requirements can be given a measure of efficiency credit when calculating the EHGSE.

Some examples of heating efficiency credits for radiant heaters are as follows:

- optimised stop   1%

- zone control   1%

For example, a flued radiant heater of rated seasonal efficiency of 84%, fitted with time control, black bulb thermostat, optimised stop and modulating burner would accrue an additional 2% heating efficiency credit which would increase it to 86%.

Combined Heat and Power (CHP) in new and existing buildings

CHP units are normally used in conjunction with boiler systems, such that the constant base load is met by the continued use of the CHP unit.

CHP systems are measured and registered in accordance with the criterion for Good Quality CHP under the Combined Heat and Power Quality Assurance scheme (CHPQA). The energy efficiency and environmental performance of the CHP scheme relative to the generation of the same amounts of heat and power by separate, alternative means is given a Quality Index number (CHP(QI)).

In order to meet the requirements of ADL2A and ADL2B the CHP plant will need to have a CHPQA QI of at least 105. Also required, is that the control system ensures the CHP unit acts as lead 'boiler' and metering is provided to detail the hours run, electricity generated and the amount of fuel supplied.

Also to meet the requirements of ADL2A and ADL2B, the CHP system needs to be sized so that no less than 45% of the total heating demand including space heating, hot water and process heating are supplied by the CHP unit unless there are overriding practical or economic constraints.

Water heaters in new and existing buildings

As well as Building Regulations there may be other overriding regulations that will need to be considered for domestic hot water systems, e.g. approved regimes in the control of legionella. The implementation of energy saving measures should not compromise the ability of the system to meet these other demands.

The Non-Domestic Building Services Compliance Guide defines the types of gas-fired systems into 2 basic types, namely:

- direct-fired systems
  (e.g. hot water storage heaters); and

- indirect-fired systems
  (e.g. boiler fed calorifiers).

In order to meet the requirements of ADL2A and ADL2B, indirect hot water systems in new and existing buildings will need to achieve EHGSE of 80% and direct-fired systems require a minimum thermal efficiency of 73% of gross CV.

In any case, the minimum controls package will need to include time and temperature control for both direct and indirect systems, whilst indirect systems will also require high limit thermostat to shut off primary flow if the system temperature is too high.

If the seasonal efficiency of a water heater does not meet the minimum required EHGSE for that type of heater, additional measures will need to be adopted by improving the control features of the system. As previously stated, any additional controls installed above the minimum requirements can be given a measure of efficiency credit when calculating the EHGSE.

Some examples of heating efficiency credits for water heaters are as follows:

- decentralisation in existing buildings only – 2%

- designing a system in accordance with manufacturers proven system sizing software – 2%

- integral combustion circuit shut-off device fitted on direct-fired heaters – 1%

- fully automatic ignition controls fitted on direct-fired heaters – 0.5%

Article 8 of the Energy Performance of Buildings Directive

Article 8 sets out the requirements under the European Council Directive 2002/91/EC, known as the Energy Performance in Buildings Directive (EPBD) for the 'regular inspection of boilers and air conditioning systems in buildings and in addition, an assessment of the heating installation in which the boilers are more than 15 years old'.

In essence the article sets out a requirement for either the regular inspection of boilers or provision of advice that may include boiler inspections.

Neither of these options places any requirement on the owners of the building to act on the report or advice given, but the intent is to persuade the owner to act if perceived benefits of undertaking such regular inspection results in energy savings.

It should be noted that the details of this article are still to be finalised, but ultimately it could mean that building occupants are advised to seek independent inspection of their boiler plant for competent assessment of energy efficient options at regular intervals.

Note: Further information on European Council Directives, including Article 8 can be obtained from the European Commission at www.ec.europa.eu

Building requirements for other areas of the UK

Other geographical areas of the UK not covered by the Building Regulations (England and Wales) i.e. Scotland, Northern Ireland and Isle of Man, have similar requirements to those already discussed in this Part. Indeed the information in the respective documentation will in some cases be identical e.g. Northern Ireland have harmonised their regulations to those of England and Wales, where appropriate.

Given this fact and although the particular documentations title or part letter for a given geographical area may be different to that of England and Wales, the information contained within those documents will be very familiar and as such will not be discussed in any detail within this manual.

For further guidance for specific requirements for a given geographical area, contact should be made with your Local Authority Building Control (LABC).

Scotland

In Scotland, local authorities administer building control, enforcing regulations made by the Scottish Ministers and approved by Scottish Parliament.

The essential purpose, as with all Building Regulations in the UK is to safeguard people in and around buildings. Added to this is the conservation of fuel and power used in buildings.

Similar to the Building Regulations for England and Wales, the Building Standards (Scotland) Regulations are subdivided into a number of parts pertaining to particular subject areas and are published in 'Technical Handbooks'.

These Technical Handbooks can be viewed at http://www.scotland.gov.uk/Topics/Built-Environment/Building/Building-standards/publications/pubtech

The Technical Handbooks particularly relevant to non-domestic gas-fired heating systems are:

- Non-domestic 3 'Environment'

- Non-domestic 4 Safety; and

- Non-domestic 6 Energy

Northern Ireland

The policy of the Building Regulations (Northern Ireland) is to achieve harmonisation with the regulations for England and Wales, where appropriate.

The regulations, like the previous counterparts for England, Wales and Scotland are subdivided into parts based on subject areas and the particular relevant parts for gas-fired heating systems are covered in the following Technical Booklets:

- DFP technical Booklet F1 – Conservation of fuel and power in dwellings

- DFP Technical Booklet F2 – Conservation of fuel and power in buildings other than dwelling

- DOE Technical Booklet K – Ventilation

- DFP Technical Booklet L – Combustion appliances and fuel storage systems.

The Technical Booklets applicable to Northern Ireland may be viewed at http://www.dfpni.gov.uk/index/buildings-energy-efficiency-buildings/building-regulations/br-technical-booklets.htm

Isle of Man

The Building Regulations (Isle of Man), like the previous regulations, are subdivided into parts based on subject areas and the particular relevant parts for gas-fired heating systems are:

- Part F – Ventilation

- Part J – Heat producing appliances

- Part L – Conservation of fuel and power

The Regulations for the Isle of Man may be viewed at http://www.gov.im/transport/planning/build/technical.xml

Competence – 3

3 – Competence

Introduction

Gas Safe Register rules for registration require individual operatives to be competent to carry out gas work. Competence in safe gas work in relation to gas fittings requires knowledge, understanding and the practical skills to carry out the work in hand in such a way as to prevent danger to life and property.

Gas Safe Register will accept that gas-fitting operatives are competent to carry out gas work if they hold valid certificates of competence to cover the areas of gas work that they intend to undertake.

For non-domestic gas work, Gas Safe Register accepts certificates of gas safety competence issued under the following schemes:

- The Nationally Accredited Certification Scheme (ACS) For Individual Gas Fitting Operatives

- A Gas Services National/Scottish Vocational Qualification (N/SVQ).

Subject to conditions that apply.

Non-domestic heating competence

Gas operatives need to have a clear understanding of the work activities that they are engaged in and importantly what certificates of competence cover those areas of work when measured against the requirements of the ACS.

The following information is provided as guidance only.

Although it covers the common non-domestic heating ACS requirements, the examples are not exhaustive and as such, gas operatives are advised that where the area of work that they are engaged in is not covered by this manual (or other manuals in the non-domestic series), to seek further assistance from an Accredited Certification Body, details of which can be found at the back of this manual under **Part 12 – References – Nationally Accredited Certification Scheme (ACS) for Individual Gas Fitting Operatives – UKAS Accredited Certification Bodies.**

The following are EXAMPLES ONLY of types of work undertaken in the non-domestic heating sector and the competencies required when measured against current ACS requirements.

Example 1 – gas-fired boilers

A gas operative who is purely involved with the installation of appliances, but does not install gas pipework or connect the appliance to the gas supply would require:

- as a minimum ICAE 1LS – limited scope operative, installation of first fix of commercial appliances.

A gas operative who is purely involved with installation of gas pipework but does not install gas appliances or connect the pipework to the gas supply would require:

*   as a minimum ICPN 1LS – limited scope operative, installation of first fix of commercial pipework.

A gas operative who is involved with the full installation, commissioning and/or maintaining of natural gas-fired boilers will require:

*   COCN1* – commercial core
    (Part A common competencies 1-6a; Part B commercial appliance common competencies 6b-11 and Part C commercial heating competencies 12-14); and

*   CIGA1 – commission, service, repair and breakdown of commercial indirect fired heating appliances and equipment

*Should a gas operative also work on Liquefied Petroleum Gas (LPG), then that operative will require a fuel changeover core, ACS code CoNGLP 1 in addition to COCN 1. COCN 1 covers installation and tightness testing of gas pipework up to 35mm internal bore and/or volume of up to but not exceeding 0.035m³.

Where gas pipework is in excess of 35mm internal bore and/or volume of greater than 0.035m³, then additional ACS assessments, namely ICPN 1 – Installation of commercial pipework in excess of 35mm and either TPCP 1 – Strength, tightness testing and purging pipework (up to 16 bars) or TPCP 1A – Strength, tightness testing and purging pipework (up to 40mbar and 1m³ in volume), will be required.

Example 2 – overhead radiant tube heaters

A gas operative who is involved with the full installation, commissioning and/or maintaining of natural gas-fired radiant tube heaters will require:

*   COCN1* – commercial core
    (Part A common competencies 1-6a; Part B commercial appliance common competencies 6b-11 and Part C commercial heating competencies 12-14); and one or more of

*   CORT1 – commission, service, repair and breakdown of commercial radiant plaque and radiant tube heaters

*Should a gas operative also work on Liquefied Petroleum Gas (LPG), then that operative will require a fuel changeover core, ACS code CoNGLP 1 in addition to COCN 1. COCN 1 covers installation and tightness testing of gas pipework up to 35mm internal bore and/or volume of up to but not exceeding 0.035m³.

Where gas pipework is in excess of 35mm internal bore and/or volume of greater than 0.035m³, then additional ACS assessments, namely ICPN 1 – Installation of commercial pipework in excess of 35mm and either TPCP 1 – Strength, tightness testing and purging pipework (up to 16 bars) or TPCP 1A – Strength, tightness testing and purging pipework (up to 40mbar and 1m³ in volume), will be required.

Example 3 – warm air heaters

A gas operative who is involved with the full installation, commissioning and/or maintaining of natural gas-fired heaters will require:

- COCN1* – commercial core
  (Part A common competencies 1-6a; Part B commercial appliance common competencies 6b-11 and Part C commercial heating competencies 12-14); and one or more of

- CIGA1 – commission, service, repair and breakdown of commercial indirect fired heating appliances and equipment

- CDGA1 – commission, service, repair and breakdown of commercial direct fired heating appliances

*Should a gas operative also work on Liquefied Petroleum Gas (LPG), then that operative will require a fuel changeover core, ACS code CoNGLP 1 in addition to COCN 1. COCN 1 covers installation and tightness testing of gas pipework up to 35mm internal bore and/or volume of up to but not exceeding 0.035m³.

Where gas pipework is in excess of 35mm internal bore and/or volume of greater than 0.035m³, then additional ACS assessments, namely ICPN 1 – Installation of commercial pipework in excess of 35mm and either TPCP 1 – Strength, tightness testing and purging pipework (up to 16 bars) or TPCP 1A – Strength, tightness testing and purging pipework (up to 40mbar and 1m³ in volume), will be required.

Example 4 – combined heat and power

A gas operative who is involved with the full installation, commissioning and/or maintaining of static gas fuelled spark ignition engines will require:

- COCN1* – commercial core
  (Part A common competencies 1-6a; Part B commercial appliance common competencies 6b-11 and Part C commercial heating competencies 12-14); and one or more of

- CGFE1 – commission, service, repair and breakdown of gas fuelled engines

*Should a gas operative also work on Liquefied Petroleum Gas (LPG), then that operative will require a fuel changeover core, ACS code CoNGLP 1 in addition to COCN 1. COCN 1 covers installation and tightness testing of gas pipework up to 35mm internal bore and/or volume of up to but not exceeding 0.035m³.

Where gas pipework is in excess of 35mm internal bore and/or volume of greater than 0.035m³, then additional ACS assessments, namely ICPN 1 – Installation of commercial pipework in excess of 35mm and either TPCP 1 – Strength, tightness testing and purging pipework (up to 16 bars) or TPCP 1A – Strength, tightness testing and purging pipework (up to 40mbar and 1m³ in volume), will be required.

Hot water heating systems – 4

4 – Hot water heating systems

Introduction

Water boilers and their various systems are one of the most common forms of gas utilisation for heating purposes in domestic, commercial and industrial properties.

There is a wide variation in sizes of hot water central heating systems that the non-domestic gas operative can encounter, from small public houses up to large commercial and industrial sites.

Note: For information relating to large domestic and small commercial sized central heating systems, see – Central Heating – Wet and Dry – from CORGI*directs* current Gas Installer Manual series (Order Ref: GID7), see Part 13 – CORGI*direct* Publications.

Information contained within GID7 is relevant to the design parameters, materials and methods for the installation of central heating systems at domestic premises, but can also be applicable to small commercial premises, for example at some public houses, small B&Bs or hotels.

On large commercial or industrial sites, where there is a requirement for space heating over a wide area, such as hospitals for example, there are likely to be significant lengths of hot water distribution pipework installed.

The associated distribution losses will often mean that medium temperature hot water systems will be required for the heating system to be effective.

This Part will review various boiler systems and their applications such as:

*   different types of hot water systems (low/medium/high temperature) used for space heating purposes

*   process water heating. In many commercial and industrial premises there are process requirements for hot water, such as cleaning pots and utensils at catering establishments or shower block facilities at industrial sites where work involves exposure to dirt-laden atmospheres etc

*   industrial steam raising applications. Whilst these systems are currently beyond the scope of ACS, in terms of operative competency requirements, there will be a requirement for operative training to an Approved Code of Practice (ACOP) standard as required by the HSWA. This Part offers an introduction to steam raising systems, where they are likely to be installed, relative advantages/disadvantages and reasons for utilising steam processes.

Types of boilers found in the commercial and industrial gas utilisation market, including installation and commissioning will be reviewed later in Part 6 – Boilers of this manual.

Part 6 considers the various types of boilers generally available, from cast iron sectional boilers, including modular systems and condensing boilers, through to steel shell type boilers and reviews the relative merits of individual boiler types and systems for any particular application.

Note: Servicing and maintenance of boilers will be reviewed in Part 10 – Servicing, maintenance and fault finding.

As previously discussed in Part 2 – Gas and associated legislation, Part L of the schedule to the Building Regulations (England and Wales) is concerned with the conservation of fuel and power in buildings.

The efficiency of heating systems fitted in buildings used for commercial and industrial purposes is covered by Part L, Approved Documents L2A (ADL2A – for new buildings) and L2B (ADL2B – for existing buildings).

These two documents give guidance on how to satisfy the energy performance requirements in non-domestic buildings.

Whilst the efficiency of the appliance itself is an important feature of these building regulations, it is also important to manage the system efficiently and minimise energy losses due to boilers coming online too early in the day during autumn or spring seasons and idling unnecessarily.

To improve energy performance of hot water heating systems in non-domestic buildings it is possible to introduce additional energy saving devices such as optimisers and compensators.

Optimisers are used to optimise start time of the boilers each day dependent on inside and outside temperatures. Compensators control the recycling of hot water in the return pipe system to satisfy room temperature before firing the boiler. That is, if there is already a high return flow water temperature when the thermostat re-initialises, the compensator will open-up a three port valve on the return flow to recycle the return hot water until return flow temperature reduces sufficiently to warrant the boiler firing.

It is also important to ensure that the boilers are correctly sized to meet the requirements of the load at any given time. It can sometimes be cost effective to install a large boiler to meet the needs of the winter demand of space heating and 'domestic' hot water, whilst a second smaller boiler (or direct water storage heater) is used purely to meet the summer demand for 'domestic' hot water use only.

Or alternatively, modular boiler systems will save energy where there is a wide variance in load requirements, not only from season to season but also throughout the demands of the day. Hotels for example, will often be unoccupied for most of the day but have heavy demands for heating and hot water in the evening and early morning.

Low Temperature Hot Water (LTHW) systems

Commercial premises such as offices, public houses, B&Bs and hotels are usually most suited to LTHW systems whereby the heat generated at the boiler is transferred to water circulating through a system of small bore distribution pipes, to a number of radiators located at strategic points along the system.

LTHW systems operate in the temperature range of 70°C to 100°C. However, the nominal working temperature will normally be a maximum of about 83°C to allow a margin of safety in the event of overheating.

Figure 4.1 illustrates a typical vented LTHW system for combined heating and domestic hot water. The system can incorporate a double pipe flow and return system, connected to the heating and hot water load with individual pumped circuits and three port valves. A programmer, possibly with integral optimiser and compensator, will be used to control heating/hot water times and a thermostat to control temperatures.

It is commonplace to incorporate thermostatic radiator valves (TRVs) on each radiator to allow variety of temperature control in different rooms.

Note: When all radiators are fitted with TRVs, a recirculating bypass facility will be required to ensure that the circulation pump does not act against a completely closed system, as would be the case if all TRVs were in the closed position.

This 'traditional' system incorporates a separate open vent pipe that is discharging over a feed and expansion (F&E) cistern, sited at the highest point in the installation.

The cistern will be supplied with cold water from the incoming water distribution main and have a separate cold feed pipe supplying water to the circulating pipe system.

Figure 4.1 Typical LTHW system

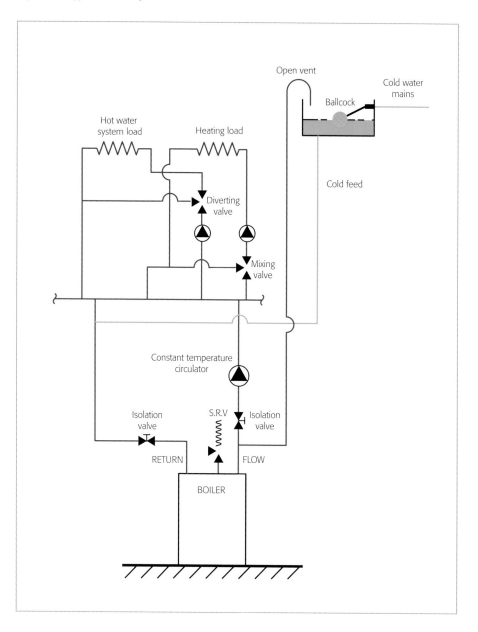

Figure 4.2 Typical LTHW system providing space heating with modular boilers

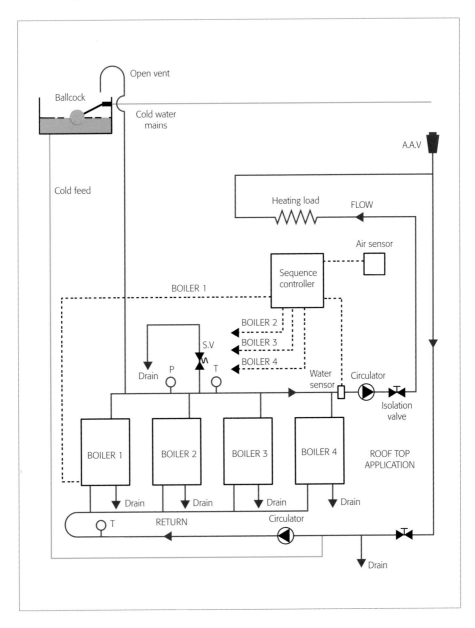

A variation of the LTHW system is shown in Figure 4.2 opposite, which shows a typical vented system for space heating incorporating modular boilers.

Modular boilers are used when the connected load is of sufficient magnitude that a single boiler of the correct size becomes costly and inefficient.

The heat demand on a typical commercial heating system can vary enormously throughout the day and from season to season. To try and match the peak demand with a single boiler will subsequently mean that there will be significant periods of time when the boiler is idling, i.e. just maintaining hot water in the body of the boiler ready to meet the demands of the variable load. The larger the boiler, the greater the level of inefficiency.

This is undesirable for two principle reasons – the excessive cost of the fuel wasted and also the unnecessary emissions of Carbon Dioxide (CO_2) to the atmosphere.

A system incorporating modular boilers, utilises a series of smaller output boilers that are systematically linked with each other. Each boiler is independent, in as much as it has its own burner, ignition and thermostat control, but the hot water flow and return pipe circuit will interconnect to each boiler.

At any given time the heat input to the system can be matched as dictated by demand, i.e. the number of boilers online at any time will increase with increase in demand. Each boiler will therefore operate at its optimum efficiency, which will reduce cycling and improve system efficiency.

If the boiler sequencer control unit then varies the start-up sequence on a regular basis, by automatically sequencing which boiler lights up first to meet initial demand, each boiler will wear evenly.

Typically a modular boiler system may have as many as six individual boilers interconnected.

Heat distribution for any type of LTHW system can be introduced to the building by a variety of mechanisms.

Traditionally, the hot water will be distributed to pressed steel or cast iron radiators. Approximately 50% of the heat dissipated by a radiator is by radiation and the remaining 50% by convection, although this will vary based on a number of factors. These include, radiator position; surface 'emissivity' (a measure of the ability of a surface to radiate energy); air movement; etc.

Natural convectors are designed to increase the convection output and can be of the cabinet type or low level skirting types. These are radiators, which incorporate additional finned surfaces to increase emissivity and thus the amount of heat given-up to convection.

Fan assisted convectors will have a much higher convective output and will traditionally be more suited to the heating of larger work areas due to increased noise levels and air movement. The fanned unit will usually incorporate fan control switches, to operate the fan when the heating element is up to operating temperature and ensure the unit does not continuously blow cold air, when the hot water circuit is not in use.

Modern commercial premises, such as office accommodation; sports facilities and club rooms are quite likely to be fitted with underfloor heating.

These systems comprise a series of heating pipe loops of flexible polybutylene, polyethylene or a composite pipe material, installed beneath the floor surface on heat reflective underlay. Each looped circuit can be controlled by a thermostatic valve at a header manifold. The flow and return header manifolds are connected to the flow and return connections on the boiler.

Underfloor heating offers some significant advantages to architects when designing a building; it will allow them to create an aesthetically pleasing building interior without unsightly central heating pipes and radiators.

Also, compared with conventional heating systems, energy savings are not uncommon, as the whole of the floor surface effectively becomes a radiator; the fabric of the building is being heated directly by the heating circuits.

Unvented hot water storage systems

The Building Regulations (England and Wales) were amended in 1985 to provide safety requirements for unvented hot water storage systems. Subsequently, Water Byelaws were also amended to override the previous restriction on hot water storage systems, which limited storage to a maximum of 15 litres.

The alterations to these Regulations mean that hot water circulating systems can now be connected directly to the water mains and do not have to be vented as described in the previous section. Instead, the hot water can be stored under pressure in a vessel incorporating expansion and temperature relief.

The obvious advantage of this arrangement is that of eliminating the need for a cistern located at high level and the associated difficulties of running the cold water feed to it. Also if there is no cistern, there is no likelihood of the water contained within it becoming contaminated.

An unvented system therefore offers a certain amount of increased flexibility.

Figure 4.3 illustrates a typical unvented LTHW system.

With an unvented system, cold water is fed from the incoming water mains into a pressurisation unit via a filter, a pressure reducing valve to control pressure at a nominal 2 bar and a non-return valve to prevent back-flow should mains pressure reduce.

An expansion vessel is provided to accommodate the increase in volume of the expansion of water under normal heating conditions.

As the water in the system is heated by the boiler, the expanded volume is absorbed by the expansion vessel. A small rise in pressure takes place, which the vessel is designed to accommodate.

An expansion relief valve automatically discharges any excess pressure if the predetermined pressure is exceeded.

In the event of system fluid loss when the system cools down, water stored in the pressurisation unit accumulator will re-charge the system pressure. If the amount of fluid loss is greater than that stored in the accumulator, a pump will activate to re-charge the system and maintain the minimum required cold fill pressure.

However, if the fluid loss is due to a major system leak and the pressure continues to fall, a system low-pressure switch will shut down the boiler and pump ensuring a fail-safe condition.

Hot water storage systems (gas-fired water storage heaters)

Where there is a space heating requirement being fulfiled by a boiler and LTHW system, it is usually practical to use the boiler to also provide the heat to a calorifier for the generation of domestic hot water.

However, if the main space heating boiler is the only source of heat for generating the hot water, at times of low heating requirements, e.g. during summer months, the main boiler must be fired just to provide hot water.

This means that the boiler will not be operating at its optimum efficiency for the whole of the summer period and will be idling for most of this time.

Figure 4.3 Typical unvented LTHW system

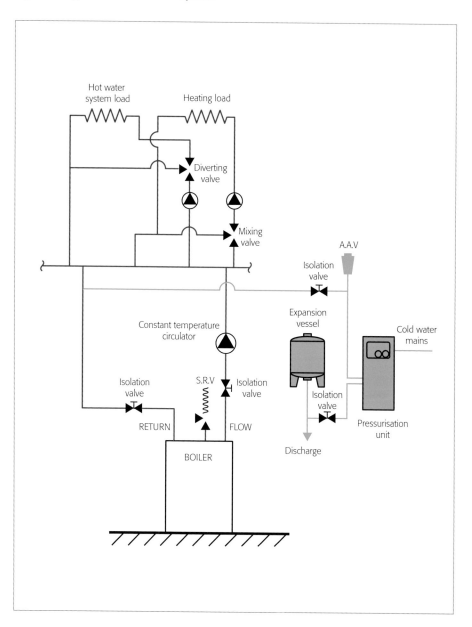

It is quite common in large commercial or industrial boiler houses to find a gas-fired hot water storage unit being utilised during summer months to provide domestic hot water. Alternatively, where the hot water requirement is the dominant load, such as at catering establishments, nursing homes or industrial processes using hot water for cleaning, it can be more efficient to heat the water directly by this method.

Although individual manufacturer's units vary in design, the storage vessel usually comprises an internally glass lined steel tank, which is under-fired using a natural draught atmospheric burner. The products of combustion (POC) fire up through single or multiple tubes within the tank to exhaust.

Each tube can be fitted with swirl baffles to increase turbulence of the combustion gases and thus increase efficiency of heat to water.

For further details on this particular type of appliance refer to **Part 6 – Boilers** in this manual.

Figure 4.4 details the basic system for a single vented water storage heater.

The individual units are available in a wide range of capacities, but can be interconnected with additional heaters to increase temperature recovery capacity.

Figure 4.5 details the basic system for two direct-fired water storage heaters, installed in tandem for increased recovery capacity.

Similar to a conventional boiler installation, it is possible to install these water heaters on an unvented system (see Figure 4.6).

Alternatively, a single water heater can be interconnected with an adjacent storage vessel to increase the volume availability where large volumes are required instantly, but longer recovery periods to reach the set temperature are allowable. Figure 4.7 illustrates such a system.

Medium/high temperature hot water (MTHW/HTHW)

Medium and high temperature hot water systems are mainly used on large 'commercial' sites such as hospitals; military bases; colleges and universities; industrial factories over widespread areas etc. Traditionally, anywhere where there is a centralised boiler plant, distributing hot water around a large widespread site for space heating purposes.

LTHW systems discussed previously in this Part use a typical temperature drop of 11°C and a maximum drop of 17°C; however, on a large distributed hot water system, distribution losses mean that temperature drops can be much higher than this.

In order to compensate for the hot water distribution losses expected of such a system, the distribution system is pressurised. Raising the pressure of the hot water raises its boiling point so that higher water temperatures can be achieved without boiling taking place.

MTHW systems operate at temperatures in the range 100°C to 120°C, whilst HTHW systems operate between 120°C and 200°C. Correspondingly, operating pressures are in the range 3bar to 11bar.

Pressurisation of the system can be achieved by various methods: a closed system incorporating an expansion vessel, steam cushion or external nitrogen pressurisation.

Where the hot water system is distributed in high rise buildings, the head of water will also result in a much higher working pressure at the base of the system. Therefore, if the boiler room is located at the base of the system, the header pressure effect is something that needs to be borne in mind when selecting the correct boiler for the application.

Figure 4.4 Typical vented water storage heater system

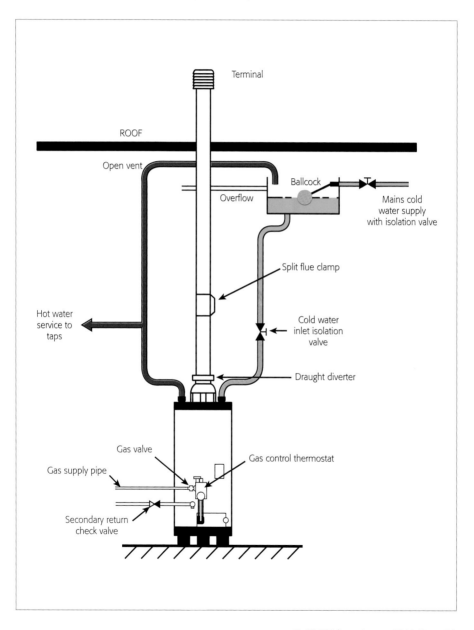

Terminal

ROOF

Open vent

Ballcock

Overflow

Mains cold water supply with isolation valve

Split flue clamp

Hot water service to taps

Cold water inlet isolation valve

Draught diverter

Gas valve

Gas control thermostat

Gas supply pipe

Secondary return check valve

Figure 4.5 Typical vented system incorporating two heaters

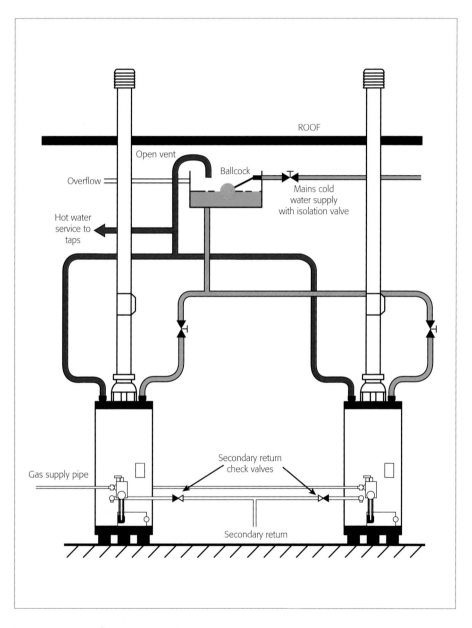

Figure 4.6 Typical unvented system incorporating two heaters

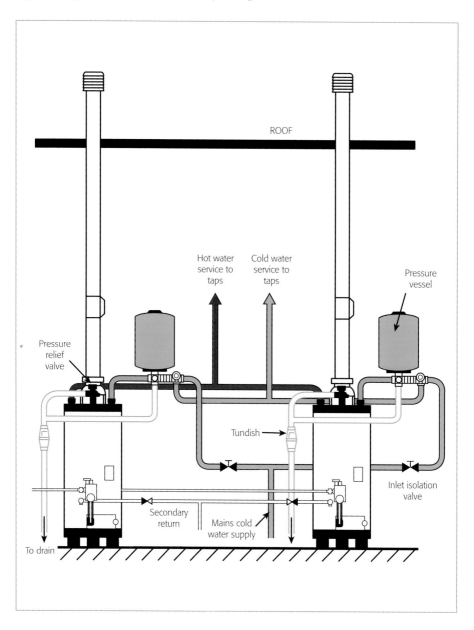

Figure 4.7 Typical vented water heater interconnected with storage vessel

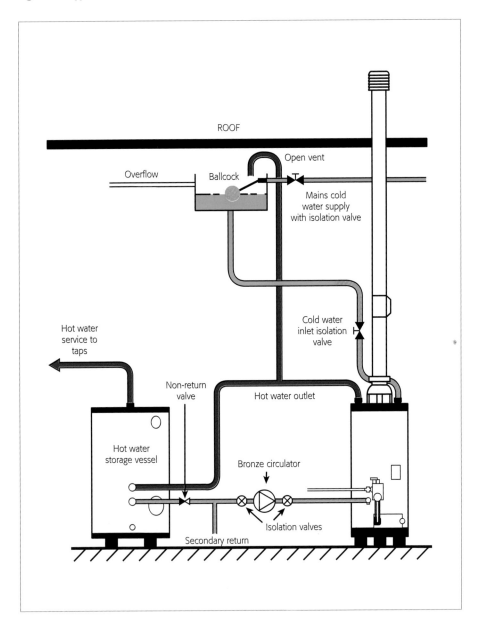

With the appropriate boiler type capable of withstanding the additional pressures, e.g. steel shell type, the boiler will be more compact and heat transfer rates will be higher, compared to LTHW systems. An advantage of MTHW/HTHW systems is the space saving in the boiler house.

Types of boiler used in this application are reviewed in **Part 6 – Boilers** of this manual.

The MTHW/HTHW system can be distributed direct to suitable heat emitters, i.e. radiators or convectors designed to operate at the greater pressure. Alternatively, the system can be used to heat point of use calorifiers, from which a secondary distribution system operating at LTHW is installed.

Steam systems

Steam systems are widely used principally for heating processes, power generation and space heating.

Steam distribution systems have historically been widely used in industry for space heating purposes only. However, when using premium fossil fuels such as NG and LPG or 35 second oil, it is now generally considered to be a costly method of providing space heating, compared in particular, with decentralised systems using warm air or radiant heating.

For details of decentralised warm air and radiant heating methods refer to **Part 7 – Overhead radiant heating** and **Part 8 – Warm air heating systems** of this manual.

Where there is a process need for steam, e.g. at hospitals; industrial laundries; food production factories; etc. which will be required for sterilisation purposes, it can prove economic to use the steam generated to additionally provide the source for space heating. Figure 4.8 illustrates a typical steam system for process and space heating and Figure 4.9, the associated steam controls.

There are some advantages in using steam systems in this way compared to HTHW systems, namely:

- they do not usually require a circulation pump, except for condensate returns

- pipework can be smaller for a given heat requirement

- steam to LTHW calorifiers/heat exchangers can be smaller

- pipe system faults/breakdowns can generally be dealt with quicker due to faster cooler times and reduced drain down requirements

- heat emitters can have higher rated outputs with steam.

There are three stages in steam generation –

1st wet steam

2nd dry saturated steam; and

3rd superheated steam

Wet steam is produced when water is raised to 100°C at atmospheric pressure (temperature will be higher for pressurised systems). The steam contains a proportion of water in suspension and the amount of energy required to evaporate water into steam at the same temperature and pressure will be approximately 2250kJ per kg (note 1kW = 1kJ/s).

Dry saturated steam is produced by raising the temperature until all water is completely vaporised. If the temperature of the steam is raised still further, superheated steam is produced.

The ratio of actual steam to wet steam is known as the 'dryness fraction', e.g. if 10kg of wet steam contains 2kg of water its dryness fraction will be 0.8, in other words only 80% of the water has been vaporised.

Figure 4.8 Typical steam system

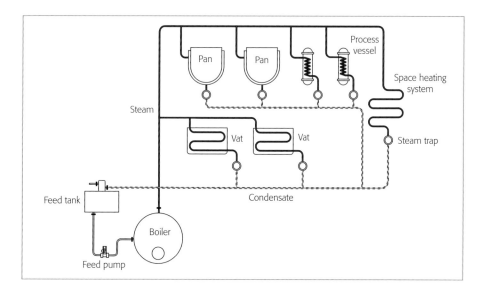

Figure 4.9 Typical steam systems controls

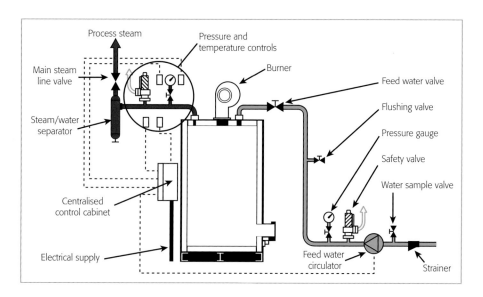

Steam boilers are denoted by their steam output capacity, which is the amount of water in kgs that can be evaporated per hour at 100°C, e.g. steam output quoted in kg/hr at 100°C.

To raise 100kg of steam per hour will require =

$$\frac{2250kJ \times 100kg}{3600secs} = 62.5kW$$

If the boiler is 80% efficient, the input rating =

$$\frac{62.5kW \times 100}{80} = 78.12kW$$

Note: For further detailed guidance on steam heating, reference can be made to The Combustion Engineering Association (see Part 12 – References) publications, entitled –

- Guide to Health and Safety in the operation of industrial boiler plant; and

- Guide to steam plant operation

Additionally, the HSE also produce a guide entitled –

- Safe management of industrial steam and hot water boilers (Code: INDG436)

Boiler locations – 5

Introduction

Commercial and industrial boilers are available in a wide variety of types and sizes, a selection of the types available are reviewed in **Part 6 – Boilers** of this manual.

Selection and location of the boiler(s) will need to take account of an equally wide number of factors, such as:

- boiler rating

- physical boiler size and weight including water content

- flueing methods – selection of most appropriate system and route to ensure effective discharge of the products of combustion (POC) to the outside atmosphere

- limitations of building – location, size, construction of available boiler/plant room

- requirements of appropriate British Standards (BS) and other relevant Codes of Practice, particularly:

  - BS 6644: 2011 'Specification for the installation and maintenance of gas-fired hot water boilers of rated inputs between 70kW (net) and 1.8MW (net) (2nd and 3rd family gases)'

  - IGE/UP/10 (Edition 3) with amendments October 2010 'Installation of flued gas appliances in industrial and commercial premises incorporating specific requirements for appliances fired by bio-fuels'

- requirements of the Clean Air Act

- requirements of Building Regulations (see **Part 2 – Gas and associated legislation** in this manual)

- local authority requirements – planning permission restrictions, e.g. for listed buildings

- Construction (Design and Management) Regulations (see **Part 2 – Gas and associated legislation** in this manual).

Note: This list should not be regarded as exhaustive.

In all cases, boiler manufacturer's information will need to be consulted and where necessary, additional guidance sought from the manufacturers as to the best practice for siting the boiler to ensure optimum performance.

To achieve complete combustion at the boiler, it is essential to ensure that it is supplied with an adequate supply of clean fresh air (ventilation). This will take into account the type of chimney fitted to the boiler and any other appliances/plant in the same area, along with any cooling requirements for the boiler.

Note: For information relating to the general ventilation requirements in non-domestic establishments, see Essential Gas Safety – Non-Domestic (Order Ref: ND1, see **Part 13 – CORGI***direct* Publications).

Information in this Part is relevant to the design parameters, materials and methods for the installation of appropriate ventilation for any particular boiler installed at non-domestic premises.

Any boiler will need to be located in a safe and secure location, ideally in a purpose built construction such as boiler house/room or plant room. Where this is not practical, if the boiler is to be located in a general factory or work space, then due recognition will need to be taken of surrounding activities and their effects on safe boiler operation, e.g. movement of personnel, vehicles (particularly fork lift trucks), chemical or flammable work processes, etc.

The floor construction on which the boiler is to be sited will need to be constructed of material that will withstand temperatures of at least 65°C and be of suitable construction to support the weight of the boiler when filled with water.

Manufacturer's instructions will give advice on the required clearance from combustible materials and the necessary clearances for safe access for maintenance purposes.

The gas supply into the boiler house/room or plant room will need to be fitted with a manual isolation valve, which is clearly identifiable and readily accessible for operation.

Automatic Isolation Valves (AIVs) have, over the years become a commonly fitted safety device in boiler rooms. Historically, this stems from incidents arising from fuel spillage or boiler fires predominantly associated with oil or coal-fired plants. Whilst they are seen in general to be a worthy safety feature, on gas pipework systems they are not without associated hazards resulting from inappropriate application.

It is of prime importance therefore, that when operatives install AIVs, or encounter them installed in boiler rooms, they consider the correct application of the system.

IGEM/UP/2 (Edition 2) 'Installation pipework on industrial and commercial premises' – Appendix 6 'Selection of a gas supply protection system' outlines the requirements for selection and installation of AIVs and identifies the potential hazard that can arise if fitted in unsuitable applications, e.g. where the pipe system feeds appliances with only permanent pilot flame safeguards or no flame safeguard at all.

Having identified the application to be correct, then the operative will also need to ensure that the type of AIV installed is suitable for the application, e.g. electrically operated and complying with BS 7461: 1991 'Specification for electrically operated automatic gas shut-off valves fitted with throughput adjusters, proof of closure switches, closed position indicator switches or gas flow control' or BS EN 161: 2011 'Automatic shut-off valves for gas burners and gas appliances'.

Note: The use of a drop weight valve or an electrical valve, which automatically restores to the open position after transient loss of power, are not generally recommended.

For certain maintenance or repair tasks there will need to be adequate facilities to permit the drain down of the system. These facilities will need to be readily accessible and also permit drain down without causing nuisance or safety hazards to personnel operating in the immediate vicinity.

Condensing boilers will involve additional considerations particularly regarding the condensate removal. The condensate disposal pipe will need to run indoors to avoid the possibility of freezing in winter. The location of the boiler can sometimes be dictated by the nearest access to soak away drainage facilities.

Also, condensing boilers can have a tendency to form a plume of water vapour from the flue terminal; consideration will therefore be required as to the effect of this pluming on the surrounding building structure, materials and neighbouring buildings.

Wherever the boiler is to be located, the designer of the system should give some consideration to the physical replacement of the boiler at some time in the future.

This is of particular importance for roof top or basement boiler rooms. Cast iron sectional boilers can usually be assembled on site if necessary, but shell boilers cannot. To retrofit a shell boiler it will often be necessary to remove part, or all of a wall or roof structure and hoist/crane the boiler into place.

Restricted locations

Individual boilers and associated water distribution systems will need to be installed in locations that are deemed suitable. In industry particularly, there will be a wide variety of applications and processes. Therefore, due consideration needs to be given to the likely environmental conditions, for example:

- corrosive or salt-laden atmospheres will obviously have an affect on metallic components, particularly burners and controls, e.g. metal plating/treatment workshops

- dusts and vapours given off from plastic forming processes. Cleaning or curing applications when passed into the atmosphere will be drawn into the combustion chamber and may subsequently produce harmful gases or explosive risk

- chlorine laden atmospheres (e.g. at swimming pools/baths). When an atmosphere is laden with chlorine in the vicinity of a permanent heat source, such as a permanent gas pilot light, the chlorine will break down to base component chemicals, including acids which will attack the metallic components of the heater, causing rapid corrosion

There have been problems in the past with industrial degreasing plants, which used chlorine based degreasing fluids. Vapours given off by the degreasing process were drawn into the appliance from the chlorine contaminated atmosphere and caused rapid deterioration of the heater. However, these fluids are no longer permitted in the UK, so gas operatives are unlikely to encounter this particular problem in future

- high velocity air movement in the vicinity of the burner will affect its performance and may prevent complete combustion from taking place

- petroleum or heavier than air flammable vapours, as a result of spillage at a vehicle workshop/garage, represents a significant risk of explosion. Any gas boiler located in such an area, will need to be sited so as not to represent an ignition source to the fuel spillage i.e. the base of the boiler will need to be mounted at least 1.8m above floor level

In addition, electricity supply cables and switchgear will need to be sited above 1.2m or appropriately protected as defined in BS EN 60079-10: 2003 'Electrical apparatus for explosive gas atmospheres. Classification of hazardous areas'

BS EN 60079-10 defines areas in distinct zones, i.e. zone 0, 1 and 2, relative to known hazards that may be present. Increasing the amount of ventilation in a hazardous area may in certain cases change the zone rating of an area and may then permit installation of gas fired appliances

IGEM has released a new document covering garages - IGEM/UP/18 'Installations in garages', which provides further details

- LPG fuelled appliances with automatic ignition or permanent pilot light cannot be located in a totally enclosed room below ground e.g. cellar or basement.

If there is any doubt as to the suitability of any appliance for a given application, always seek advice from the appliance manufacturer(s).

Most manufacturers will prefer to give, often free advice in order to ensure the boiler is appropriate for the application, rather than risk the possibility of inheriting bad publicity resulting from unsuitable/poor installations.

Requirements for boiler houses

A boiler house is defined by BS 6644 as a 'dedicated building for the installation of boilers and ancillary boiler plant' and likewise, a boiler room is a 'dedicated room within a building for the installation of boilers and ancillary boiler plant'.

Any boiler house/room construction needs to comply with The Building Regulations (England and Wales), The Building Standards (Scotland) Regulations or The Building Regulations (Northern Ireland) dependant on the geographical area of the boiler house/room.

For further information regarding Building Regulations refer to **Part 2 – Gas and associated legislation** in this manual.

Note: Many 'existing' boiler houses/rooms may have been built before the 1980s, whereby the use of Asbestos Containing Materials (ACMs) was wide spread. As such ACMs maybe encountered in today's working environment.

ACMs were used for their insulation properties and therefore can be found as pipe lagging, fire separation barriers, wall and door panels, ceiling and floor tiles and many more applications besides.

Responsible person(s) for the premises should be aware if asbestos is indeed present as they have a legal duty to survey the premises and to inform persons of its location and condition before any work commences.

However, should a gas operative come across ACMs whilst doing their work, they must STOP work and report their findings back to the responsible person for the premises (and where appropriate their employer) so that the appropriate course of action can be taken.

Boiler houses/rooms should not be located at low lying areas where there are risks of flooding, or if there is no alternative, suitable protection will need to be provided against the effects of flooding.

A suitable means of access to the boiler house/room will need to be provided and it is preferable that this access is from outside the building. Any means of exit from the boiler house/room will need to be through a door, which is openable from within the boiler house/room without the use of a key.

The door itself needs to be hinged, to open outwards in the direction of any escape route and it will be necessary for the source of escape in the event of an emergency to be located no more than 12m from any point within the boiler house/room. Also, adequate lighting, including emergency lighting needs to be provided to ensure ease of escape.

If the boiler house/room houses boilers operating on LPG, any doors (or windows) are not allowed to open in to any adjacent room, which is wholly below ground.

Temperatures within boiler house/room's needs to be controlled by provision of adequate ventilation so that the temperature does not exceed:

- 25°C at floor level
  (100mm above the floor level)

- 32°C at mid-level
  (approximately 1.5m above floor level)

- 40°C at ceiling level
  (100mm below ceiling level)

It will be necessary to consider the standard of gas and appliance installation with respect to the structure of the building. It is a requirement that the building structure is not adversely affected in the event of a failure of containment of the fuel or failure of any control system.

For example the resultant explosion in the event of a major gas leak could lead to catastrophic failure of the boiler house/room structure. The effect of that damage will need to be minimised in terms of knock on effect to the rest of the building or adjacent properties.

The type of boiler room locations where this can represent a significant problem include, but are not limited to:

- basement boiler/plant rooms

- fully enclosed reinforced concrete boiler/plant rooms

- mid floor level boiler/plant rooms in high-rise buildings.

To minimise the risk of such an event arising, there are a number of precautionary measures to consider, which can include:

- pipework and associated control components could be installed and tested to a higher specification than would otherwise be necessary, e.g. install heavy grade steel pipe with all joints welded, install integrally flanged control components (Safety Shut Off Valves (SSOVs) for instance) and apply a strength/tightness test at a slightly higher pressure and/or for a longer period than would normally be required

- provide adequate explosion relief by installing blow off panels (or a blow off roof for single storey structures) whereby pressure relief is automatically achieved in the event of an explosion. Any such relief panel will need a suitable method of 'flexible' restraint and, when operated, will need to be directed towards safe areas away from those normally habited

- ensuring all ventilation systems operate safely and in the event of mechanical failure the system reverts to standby or shut down

- regular routine visits by competent persons to inspect and check operation of safety systems

- install gas/carbon monoxide detector alarms linked to the incoming AIV to isolate the gas feed in the event of the detector operating. For information on AIVs see **Introduction** in this Part.

Finally, it is a requirement that all relevant design information and risk analysis relating to the safety of the boiler location is included with the health and safety file normally left with the responsible person.

Requirements for plant rooms

In terms of building structure, location and boiler requirements, all items considered previously for boiler rooms will usually apply to plant rooms.

A plant room is defined as a 'room in a building which houses plant and machinery'.

By definition then, a plant room will house other items of plant besides the boiler, e.g. refrigeration plant; water treatment plant; power generation plant; mechanical air distribution/ventilation ductwork systems; or other fuelled equipment, etc.

Where the boiler is located in a plant room with other fuelled plant or atmosphere consuming plant, a risk analysis will need to be carried out to establish the safety and suitability of the equipment running alongside the boiler.

It will be necessary to ensure that the fresh air intake is adequate for all items of plant and that POC are all correctly discharged to the atmosphere.

Inevitably, where other items of plant are located in the same area as a boiler room, some consideration should be given to the heat generated by that item of plant and the boiler together (the temperature will need to be controlled as previously mentioned for boiler houses/rooms).

Any heat gain cannot be allowed to affect the performance of either item of plant. In particular the electronic control equipment, such as burner sequencers, fitted to boilers could be affected if ambient temperatures in the vicinity of the unit rise to 40°C or above.

Requirements for rooftop installations

Locating boilers in the basement or ground floor area of multi-storey buildings can create problems with respect to the chimney installation and choice of boiler based on the header pressure effect.

The chimney (flue pipe) will often need to run externally, which can be aesthetically offensive if there are long lengths visible and can sometimes create difficulties with respect to planning permission, where this is a requirement. Alternatively, running the chimney internally over the height of a high-rise building can also be so problematic as to be impractical.

Although there are other flueing systems to consider, such as fan flue dilution, these too have relative disadvantages; a fan flue dilution system will require ductwork, safety interlocks and careful consideration as to the discharging of the POC.

A rooftop boiler house will eliminate these flueing difficulties.

The header pressure effect caused by the water in the system at a high-rise building, considered in **Part 4 – Hot water heating systems** in this manual, will mean that the boiler located at the basement or ground floor will need to be capable of withstanding this extra pressure.

Often this can dictate the need for a shell type boiler. Whereas, if the boilers are located on the rooftop, the header pressure will not affect the choice of boiler and often conventional 'cheaper' boilers can be utilised.

Whilst there are benefits to siting boilers for multi-storey buildings on the rooftop, there are some additional considerations:

- the rooftop boiler house/room needs to be designed so that any water leakage from the boiler(s) does not affect the building below it

- there will also need to be an appropriate drain down facility, again to ensure the building below is not affected when this work is carried out

- access arrangements will need to be reviewed where there is the need to carry heavy replacement boiler components e.g. cast iron boiler sections

- access arrangements for complete replacement boilers will need to be reviewed. If it is not possible to hoist the boilers up through the building, e.g. up lift shafts, stair wells or service ducts, it may be necessary to hire expensive crane facilities, which in itself, can be problematic in built-up urban areas

- rooftop boiler houses can be subjected to severe weather conditions, e.g. lightning; therefore any control system will need to accommodate additional safety features to protect the boiler(s) in such circumstances.

Requirements for balanced compartments

A balanced compartment provides a method where an open-flued appliance (Type B) can be installed in a room-sealed situation. Figure 5.1 illustrates the concept of a balanced compartment

In principle, by taking air for combustion and ventilation through air intakes sited at the top of the boiler room, passing down through ducts formed in the sides of the chimney system to an enclosed boiler room below, a balanced condition is created.

Figure 5.1 Balanced compartment

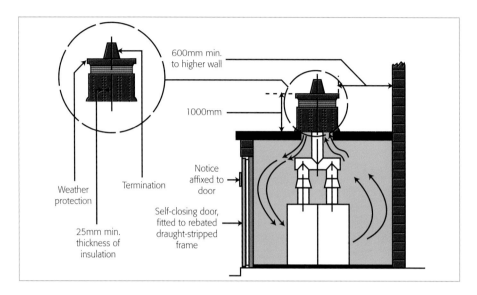

The flue gases rising directly from the boiler(s) are contained in an insulated twin wall chimney passing through the centre of the balanced flue arrangement.

Any wind movement, no matter how slight, is encapsulated by the air intakes on the windward side of the chimney and this air is conveyed to the boiler house below. Any excess wind pressure is vented on the leeward side of the system, leaving the optimum operation of the appliance unaffected by wind turbulence.

The balanced compartment system will be purpose designed for any given application, it must be recognised therefore that this is a specialist area and expert advice will need to be sought wherever it is proposed to install this type of system.

This process will also involve boiler manufacturers to ensure that the boiler operates correctly, including removal of POC and boiler room operating temperatures.

Where the flue termination is within 2.5m of an adjacent structure, the termination will need to rise at least 1m above that adjacent structure.

Figures 5.2a and 5.2b shows examples of balanced compartment terminations.

To ensure the system operates correctly it is important to ensure the boiler room itself is sealed, apart from the terminal arrangement. Any access door to the boiler room will need to be self-closing and fit tightly to the frame including a draught sealing strip. A notice will need to be attached to the door advising that it will normally have to be kept closed.

If the access door opens to an internal habitable space, the door will need to be fitted with an interlock switch to ensure that the boiler is shut down when the door is open. Where the boiler is in a plant room however, other installed equipment may need regular access; in these circumstances it is allowable to incorporate up to a 30 second time delay into the door switch operation.

Figure 5.2a Example of balanced compartment roof terminal.

Figure 5.2b Example of balanced compartment roof terminal.

The only other means of opening into the compartment is that of a 60cm² minimum fresh air intake which is required if the boilers fire on LPG. This will ensure that, in the event of a small gas escape, there is no build-up of an explosive gas mixture.

The advantages of a balanced compartment are:

- eliminates the effect of downdraught associated with normal open-flued systems

- can operate on just 1m chimney height as a vertical balanced flue

- any single storey building/room can be used as a boiler house/room

- a balanced compartment can be located where there are no immediate outside walls, eliminating the need for expensive duct systems to provide fresh air for combustion and unsightly chimneys

- no low level ventilation required

- no moving parts that would be associated with alternative systems requiring mechanical exhaust/ventilation.

Other installations

This refers to boiler installations installed elsewhere other than in a boiler house, boiler room or plant room.

Typically, it is quite common to find office accommodation located as part of the integral building structure of a manufacturing factory or commercial warehouse outlet. The boiler installation for the office complex in these situations, can be located in the factory or warehouse space rather than installed in a purpose built boiler house/room.

In commercial kitchens apart from the cooking processes, there will be a need for large amounts of hot water for cleaning purposes. To supply the volume of hot water needed, a direct-fired storage water heater could be located in the general kitchen area.

Any boiler or water heater will need to be located in a safe and secure location where it will not cause a hazard to people working in the immediate vicinity and also where it cannot be damaged by other work activities, such as movement of fork lift trucks. If necessary, the boiler location area may need to be surrounded by an appropriate barrier.

One particular problem associated with siting a boiler in a general workspace is maintaining the manufacturers recommended clearances at all times. Staff employed in the general location of the boiler can often use the available space around the boiler for storage purposes. It is important to stress to the occupants the need for appropriate clearances; it is often prudent therefore, to ensure suitable notices are displayed to this effect.

Where the boiler is located in a large factory or warehouse space having a designed building volume air change rate in excess of 0.5 per hour additional ventilation is not normally required. However, it is important to take account of any manufacturing or cooking process that requires the use of an air extraction system, e.g. welding bays, paint spray booths, kitchens etc. These extraction requirements can create a negative pressure atmosphere in the room, which will have an adverse effect on the boiler flue performance.

To achieve complete combustion at the boiler, it is essential to ensure that it is supplied with an adequate supply of clean, fresh air (ventilation). This will take into account the type of flue fitted to the boiler and any other appliances/plant in the same area, along with any cooling requirements for the boiler.

Note: For information relating to the general ventilation requirements in non-domestic establishments, see Essential Gas Safety – Non-Domestic (Order Ref: ND1, see Part 13 – CORGI Publications).

Boilers – 6

6 – Boilers

General

Boilers are probably the most common type of gas-fired equipment in the domestic, commercial and industrial sectors of the UK gas utilisation market.

They are frequently used throughout all market sectors in the generation of hot water for space heating purposes (central heating) and additionally in the commercial sectors for bulk hot water uses. Industrially, boilers are also used in the production of steam for a variety of processes and power generation.

This Part will review the common types of commercial and industrial boilers available for space heating and hot water storage and will discuss the various merits of each type.

Whilst steam raising boilers are usually beyond the scope of the ACS, in terms of operative competency requirements, this Part offers an introduction to the types of steam raising boilers that can be commonly found on some industrial sites other than for power generation.

Note: For information relating to domestic and small commercial sized central heating boilers, see Central Heating – Wet and Dry (Order Ref: GID7) and/or the Wet Central Heating Design Guide (Order Ref: WCH1), see Part 13 – CORGI*direct* Publications).

Information contained within GID7 and WCH1 is relevant to the design parameters, materials and methods for the installation of central heating systems at domestic premises, but can also be applicable to small commercial premises, for example at some public houses, small B&Bs or hotels.

Boilers for heating systems

The type of boiler utilised for any particular application will depend upon a number of factors:

*   connected load, e.g. number of radiators and size of the pipe distribution system

*   load factor, e.g. variation in daily and seasonal demand

*   location of the boiler, e.g. basement or roof-top

*   what type of wet system, e.g. whether the system is LTHW, MTHW or HTHW (see **Part 4 – Hot water heating systems** in this manual)

*   previous history of fuel firing, e.g. whether the boiler previously fired on coal and/or oil

*   requirements of the Building Regulations pursuant to the geographical area. The Building Regulations are covered in greater detail in **Part 2 – Gas and associated legislation** in this manual

*   flue system, e.g. selection of the boiler can be dictated by system/route of the most appropriate flue system to ensure effective discharge of combustion products to the outside atmosphere

*   limitations of building, e.g. location, size construction of available boiler/plant room.

Note: This list should not be regarded as exhaustive.

Cast iron sectional

Each commercial/industrial boiler manufacturer will produce a range of boilers designed to meet the needs of a wide range of applications.

The most basic of which is a cast iron sectional natural draught boiler with atmospheric burner and thermo-electric flame supervision.

The atmospheric burner is located at the bottom of the boiler where gas is fed through an injector, which entrains about 50% – 60% of the air required for complete combustion (primary aeration).

The remaining 40% – 50% of the air required is gathered from around the outside of the combustion chamber and used at the point of combustion (secondary aeration), where the flames fire up through the sections of the heat exchanger.

The cast iron sections of the combustion chamber/heat exchanger will be assembled either vertically or layered horizontally dependent on the manufacturer's individual design.

This type of burner system will require a comparatively large combustion chamber to ensure that the volume of air required for complete combustion can be drawn in.

Typically, the exit flue gas temperature of this type of boiler will be about 220°C to 260°C, producing efficiencies approaching 80% based on gross calorific value (CV) of the fuel. However, modern boilers are designed to achieve slightly lower exit temperatures and therefore higher efficiency, i.e. 80% plus.

This type of boiler is generally available with various ignition and burner control options, to automate the light-up process and facilitate additional options such as an exhaust fan, to overcome pressure resistance of difficult chimney installations or secondary heat exchangers on condensing units.

Larger versions of the cast iron sectional boiler will usually utilise forced draught burner systems, firing horizontally, into the vertically assembled sections of the combustion chamber/heat exchanger.

The forced draught burner will draw in 100% of the air it needs for complete combustion to the burner head enabling a comparatively smaller combustion chamber than that required for atmospheric burners.

Increased output capacity of cast iron sectional boilers is achieved by adding extra sections, which will increase the water carrying capacity of the boiler and thus, heat transferred to the water.

Cast iron sectional boilers are known to have long working lives, but in principle, if a section starts to leak at some point, then the boiler can be disassembled and the section individually replaced rather than replacing the whole boiler.

However, it must be borne in mind that the process of replacing a section is labour intensive and therefore costly. Unless the boiler is relatively new, the likelihood will be that during reassembly, the older boiler sections will be placed under renewed stress with the possible outcome that they too will subsequently fail.

The efficiency of the larger cast iron sectional boilers will again be in the order of 80% based on gross CV for modern type boilers. Older boilers of this type will have been traditionally designed for coal firing and will often have been converted from coal-oil-gas over the years and the efficiency of these is likely to be about 75%. However, in some cases it is possible to retrofit swirl plates into the flue pass of the heat exchanger, which can improve efficiency by approximately 2%.

Modular

The space heating load can vary significantly throughout the day at some sites, particularly at hotels for instance, so the ability to fire the requisite number of boilers to just meet the demand at any time offers significant energy savings, as the boilers are not idling at times of minimum demand.

A modular boiler system can comprise of a number of small sized cast iron sectional boilers, grouped together to meet the total maximum demand, but also give the flexibility to reduce the number of boilers firing at any time and match turn down in demand as required.

Some manufacturers of cast iron sectional boilers offer individual boiler units with automatic ignition and high/low burner control, thereby increasing the amount of variability for turn down ratio.

Other types of boiler used for modular systems include the high efficiency standard compact module type with pre-mix burners and low water content heat exchanger constructed from finned copper tubes.

This type of boiler is available in limited capacity, individual modules of 50kW, 100kW, 200kW and 600kW.

The concept is based on like type compact modules that can bolt together and build into a single boiler unit. They can be stacked by bolting together above and alongside each other resulting in a high kW boiler output per square metre of floor area occupied.

The compact design and lightweight structure makes them ideally suited for the modern compact plant room.

Steel shell

Cast iron sectional boilers are available up to about 1.5MW heat output to water. For higher boiler output capacities and generally above 500kW, it can become more economic to consider a steel shell boiler.

Unlike cast iron sectional boilers, which can be built with a varying number of sections to increase output capacity, steel shell boilers are prefabricated as a single unit, with each unit fabricated to meet a finite output range. Steel shell boilers are manufactured up to about 6MW output to water capacity.

Because the shell boiler is a welded steel construction, it is capable of withstanding higher water pressures and therefore is extensively used for MTHW and HTHW systems, or systems with high water head pressures. For instance, on high-rise buildings where the boiler is located on the ground floor or in the basement.

The forced draught burner fitted at the front of the boiler fires horizontally into the combustion chamber with subsequent passes of the flue products through the boiler, transferring heat to the water to maximise the rate of transfer and thus boiler efficiency.

Although designs vary from manufacturer to manufacturer, the compact internal design of the boilers ensures good circulation and maximum heat transfer for minimum occupied space of the boiler. Shell boilers, can be as much as 50% smaller than a cast iron sectional boiler of the same output capacity.

However, the compact design usually means higher resistance through the flue-ways, which will often require a higher gas burner pressure than would normally be available from the distribution supply.

If elevated pressure at the primary meter is not available from the gas supplier, it will normally be necessary to install a localised gas booster, adjacent to the boiler, to lift pressure to that required to overcome boiler resistance.

Note: Further guidance on gas boosters can be found in Essential Gas Safety – Non-Domestic (Order Ref: ND1, see Part 13 – CORGI*direct* Publications).

Boilers for hot water systems

In industry and commerce there are many 'domestic' and process requirements for hot water, such as:

*   general hand wash facilities in toilet blocks

*   shower facilities at sports centres, swimming pools etc.

*   shower facilities at process factories of dirty atmospheres, e.g. mines and quarries

*   catering kitchens, for washing dirty pans and utensils etc.

- commercial laundries and laundrettes

- hairdressers and beauty salons

- rest and convalescent homes

- a wide range of industrial cleaning processes and product treatments

- swimming pools.

The main problem of most hot water applications is that demand is instantaneous. So traditionally in the past, the main method of heating the hot water would be to use a bulk storage calorifier, heated indirectly from the flow and return circuit of a conventional boiler.

Using this method, the slow heat-up response time of the relatively inefficient boiler system, was compensated for by having a sufficiently large or oversized storage vessel containing hot water to ensure the maximum demand could always be met.

Where there is also a large space heating demand, this method of heating hot water can prove satisfactory, as the main boilers will be firing most of the time to meet the space heating demand anyway. However, where the demand is mainly for hot water only, then the traditional method of heating will be costly and inefficient.

Direct-fired storage water heaters offer significant advantages by combining a relatively small hot water storage vessel, which is directly fired to provide a rapid heat recovery.

These units generally comprise of a vertical glass lined welded steel cylinder with flue tubes passing through it and usually an atmospheric burner firing underneath; although some manufacturer's produce forced draught versions.

The nominal storage capacity is usually between 280 and 380 litres; the nominal storage temperature is 65°C to eliminate bacteria causing Legionnaires Disease.

The continuous output capacity of the heaters is quoted in terms of litres per hour recovered through a specified temperature lift, e.g. 275 litres/hr based on 44°C temperature lift.

To increase output capacity of the individual water heaters, the gas burner size and/or pressure is increased accordingly.

These water heaters are available in a range of output capacities from small domestic sized units, up to a nominal maximum capacity of about 2000 litres/hr recovered through 44°C for the largest individual water heater with atmospheric burner and 3200 litres/hr for forced draught burner. If a greater output capacity is required, a number of the units can be interconnected as required to give the most flexible and energy efficient arrangement.

Alternatively, where the demand priority is for increased volume availability rather than recovery rate, it is possible to interlink a water heater unit (or units) with a storage vessel.

Whilst this type of storage water heater is ideally suited for domestic hot water applications such as shower and wash facilities, different hot water applications, such as for leisure centres, swimming pools and certain industrial processes, require a different type of high efficiency rapid recovery unit. This type of application requires a water heater with a high recovery capacity but minimal storage.

The heaters most suited for this application comprise of individual units with low water content copper or stainless steel finned tube heat exchangers that can be built-up as modules to increase capacity. The capacity range of this type of individual heater can be from 1000 litres per hour to 20,000 litres per hour.

Boilers for steam raising

Boilers used for large steam systems up to 50,000kg/hr are generally multi-pass steel shell boilers; the same as for MTHW and HTHW systems.

However, many steam raising requirements for smaller industrial, pharmaceutical, medical food and drinks applications where there is a peak demand for steam over relatively short periods of time, can be met using the smaller packaged type steam boiler.

This packaged boiler comprises a vertical welded heavy gauge steel cylinder with top mounted burner.

The burner is designed to produce a swirling flame for high efficiency combustion and low Nitrogen Oxide (NOx) levels, which fires down into the vertical chamber contained within the pressure vessel.

The 'spinning' flame reverses at the base of the chamber and passes through the secondary heating space between the pressure vessel and the external insulated jacket.

The burner, ancillary equipment and controls are all preassembled by the manufacturer and supplied to site as a package. These boilers are generally available in various capacities up to an output rating of about 1000 kg/h.

Whilst these systems are usually beyond the scope of ACS, in terms of operative competency requirements, there will be a requirement for operative training to an ACOP standard as required by the HSWA.

Types of hot water boilers for heating systems

Cast-iron sectional boiler – atmospheric burner

This type of boiler is the most commonly used commercial boiler for LTHW systems in the UK.

The boiler comprises of a preassembled group of vertical or horizontally aligned cast iron sections, which vary dependent upon individual manufacturer design. The numbers of individual sections are matched with particular atmospheric burner configurations to provide a range of boiler availability between approximately 40kW to 120kW.

For increased heat output capacities some manufacturers promote the interconnecting of several single boiler units together to form a modular boiler system. For further detail on modular boilers refer to **Modular boilers** in this Part.

To improve efficiency of the boiler, the body of the assembled sections is insulated with a foil backed fibreglass blanket.

The particular boiler assembly featured in the Figure 6.1 requires the manufacturers' recommended downdraught diverter to be fitted on the primary flue exit socket.

It should be noted that some manufacturers' boiler designs incorporate the downdraught diverter integral within the boiler, this is a point worth considering if the height of the boiler room is restricted.

The atmospheric burner located beneath the boiler castings can have a simple permanent pilot and thermo-electric flame supervision on the most basic models, incorporating such safety features as overheat detectors connected via the interrupted thermocouple.

Figure 6.1 Typical cast iron sectional boiler (atmospheric burner)

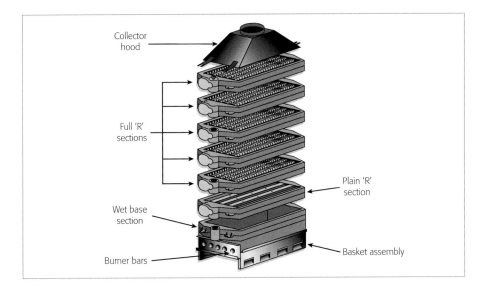

Collector hood

Full 'R' sections

Wet base section

Burner bars

Plain 'R' section

Basket assembly

More advanced models feature fully automatic burner ignition, with either a spark ignition or hot surface ignition element that reaches temperatures in the order of 1300°C and flame ionisation (flame rectification) flame supervision.

These burner sequencing units can be interlocked with other control devices such as mechanical ventilation/exhaust interlocks and automatic flue dampers.

Many of the burner systems now fitted to these boilers incorporate two-stage operation, i.e. high/low operation.

Atmospheric boilers draw in air for combustion from around the base of the boiler, aided by the pull of the natural draught flue. The flue will be sized to meet maximum load conditions so when the burner modulates to low-fire position there will be excess combustion air, which tends to reduce the efficiency of the boiler at this position.

Some manufacturers now incorporate an automatic two-stage damper in the flue exit on the boiler, to increase the efficiency of the boiler when it is firing at low-fire position.

This damper closes when the burner is off to minimise standby losses from the boiler, fully opens when the burner is on high-fire and throttles back to restrict the pull on the flue and hence the amount of air drawn into the combustion chamber when on low-fire.

The damper mechanism will be fully interlocked with the burner so that the burner cannot fire with the damper restricted or closed.

All boilers will be fitted with an overheat thermostat set within the nominal range 90°C to 110°C, to shut down the boiler in the event of an overheat condition arising.

Figure 6.2 Typical cast iron sectional boiler (forced draught burner)

Boilers with two stage burner control will be fitted with dual stage temperature control within the nominal range 65°C to 90°C, which will need to be set to provide at least a 10°C differential between high limit and normal control temperature.

Cast iron sectional boiler – forced draught

This traditional design of boiler has been used for many years for larger LTHW systems and comprises a variable number of sections assembled vertically together with the forced draught burner firing horizontally into the combustion chamber (see Figure 6.2).

The boilers are usually assembled on site and the number of sections will be increased to give more heat output to the water. Various manufacturers produce a range of different boilers to give heat output to water availability from approximately 27kW up to 1.5MW.

The sections are usually designed to give three passes of the combustion gases for maximum heat transfer. Figure 6.3 illustrates the three passes of combustion gases for a typical, traditional cast iron sectional boiler.

To improve efficiency of the boiler the body of the assembled sections is insulated with a foil backed fibreglass blanket.

The flue passes of most modern cast iron sections incorporate moulded stubs or finned projections to produce turbulent flow of the gases and maximise heat transfer to the water.

On older traditional boilers it is usually possible to retrofit swirl plates to achieve the same objective.

Each boiler manufacturer will usually offer a choice of different forced draught burner, all of which have been specifically matched to suit the requirements of the particular boiler.

Figure 6.3 Three passes of combustion gas

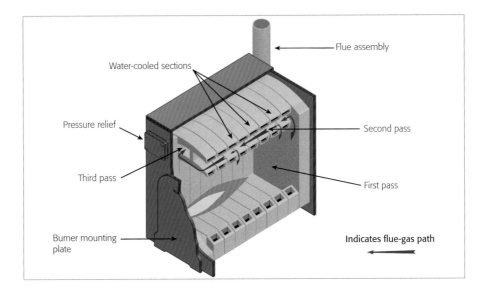

Flue assembly

Water-cooled sections

Pressure relief

Second pass

Third pass

First pass

Burner mounting plate

Indicates flue-gas path

At the lower end of the heating range the burner is likely to feature a basic on/off mode, but as the boiler capacity increases so the burner fitted will incorporate additional features like high/low operation or fully modulating control for greater efficiency and improve temperature control.

Boilers will be fitted with temperature control in a range from 35°C up to 90°C and an overheat thermostat set at approximately 110°C to shut down the boiler in the event of an overheat condition arising.

For boilers rated in excess of 1.2MW, the burner fitted will need to incorporate the additional safety feature of automatic valve proving prior to any attempt at ignition to comply with the automatic burner standard BS EN 676.

Note: For further information regarding automatic valve proving systems, reference can be made to Essential Gas Safety – Non-Domestic – Burners and their control systems (Order Ref: ND1, see Part 13 – CORGI*direct* Publications).

Condensing boilers

To improve energy performance of heating systems in non-domestic buildings, it is possible to install boilers that can operate in condensing mode.

Figure 6.4 illustrates a traditional condensing boiler arrangement whereby the combustion products generated at the burner sited below the primary heat exchanger, rise through the primary heat exchanger and then pass through an additional 'secondary' heat exchanger.

Figure 6.4 Exploded view of a typical condensing boiler

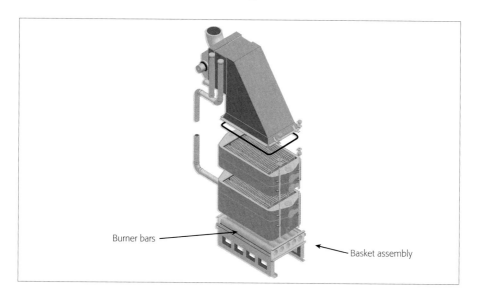

Burner bars

Basket assembly

An exhaust fan is fitted at the outlet of the secondary heat exchanger to ventilate the combustion chamber and pressurize the flue.

An alternative boiler system currently available features the primary heat exchanger sited above the secondary heat exchanger, which is positioned at the base of the boiler assembly.

A fully modulating burner, fires down through the primary heat exchanger to maximise heat recovery from the flue gases in the secondary heat exchanger.

By passing the combustion products through this secondary heat exchanger the temperature of the gases will drop below their dew point (55°C nominally) such that the latent heat is extracted from the water vapour. In this way a boiler efficiency of approximately 92% based on gross CV is achieved.

Modular boilers – multiple boilers with individual combustion chamber

These are standard or condensing cast iron sectional boilers interconnected to form an infinitely variable means of meeting any particular load requirements. Operating efficiency is maximised by firing the minimum number of boilers to match the system load at any time (see Figure 6.5).

The relatively small size of the unit makes transportation and installation easier, reducing the need for special access and time consuming assembly of the sections on site.

Another advantage is that of greater flexibility when carrying out repair and maintenance work; the heating system can be maintained, with one boiler being taken off line one at a time, whilst the others within the modular system continue to meet demand.

Figure 6.5 Typical cast iron sectional boilers interconnected as a modular system

Some manufacturers of modular boilers supply prefabricated modular pipework kits, which include:

- single pipe low loss header section

- individual boiler circuit pumps

- all necessary isolation valves

- non-return valve

- three way vent cock

- flexible connector

- gas header

- flow and return sensing manifold for connection of safety valve, flow switch, open vent and temperature gauge.

The installation of the modular pipework kit will permit the boiler sequence controller to maintain the minimum number of boilers online to meet the required load whilst isolating water flow to and from any non-firing boilers.

Without this facility, water will flow through all boilers at all times even during periods of low load when they are not firing. In this situation the non-firing boilers will effectively become an unwanted radiator load on the system and reduce system efficiency.

As many as six boilers can generally be grouped together into one manifold, any more than this will necessitate splitting the total number into smaller groups, e.g. eight boilers could be grouped into two manifolds of four.

Modular boilers will also require particular attention to be paid to the flueing arrangement.

For further guidance regarding commercial chimney systems, reference can be made to Essential Gas Safety – Non-Domestic – Chimney standards (Order Ref: ND1, see **Part 13 – CORGI*direct* Publications**).

Figure 6.6 Typical high efficiency modular boiler units

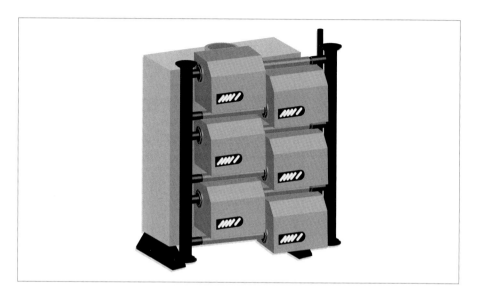

Modular boilers – multiple boilers with modular combustion chamber

This type of boiler is available in limited capacity individual modules of 50kW, 100kW, 200kW and 400kW (the 400kW unit comprises of 2 x 200kW).

The concept is based on a number of compact modules that can bolt together and build into a single boiler unit. They can be stacked by bolting together above and alongside each other resulting in a high kW boiler output per square metre of floor area occupied (see Figure 6.6).

The principle of using the boilers as modules remains the same as for conventional cast iron sectional boilers. However, the benefit of this type of boiler arrangement is its improved efficiency and reduced occupied floor area.

A typical boiler of this type comprises of a low water content heat exchanger consisting of extruded copper or aluminium alloy finned tubes in a circular arrangement around the gas burner.

The copper/aluminium tubes are expanded into cast iron headers, onto which cast iron cover plates are bolted. Figure 6.7 illustrates the typical arrangement.

The burner is a perforated stainless steel cylinder producing a flame all around its periphery. The gas pressure is controlled by a zero pressure regulator, which has an impulse tube from the positive air pressure produced by the burner combustion fan; this will ensure gas and air remain in the correct proportion for complete combustion.

The hot POC flow radially outwards around the finned tubes of the heat exchanger, giving up the heat to the water in the tubes.

Figure 6.7 Typical heat exchanger module

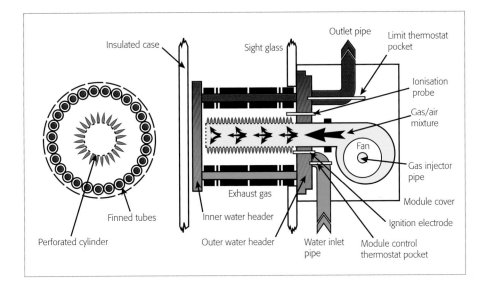

Unlike system requirements of cast iron sectional modular boiler systems, water flows through all the modules on the reverse return principle. As the load is satisfied the upper boiler modules are switched off first, then the middle and lastly the lower module. In this way, water flowing through the upper modules is kept warm by the exhaust gases from the lower modules, which are still firing.

Steel shell – single pass fire tube

These boilers are likely to be found in older established dry cleaners or in factories where clothing manufacture takes place. Typically they are used for creating the steam required for ironing processes.

They represent older boiler types with a single set of fire tubes, whereby the burner fires into the tubes at one end and combustion products exhaust to the flue at the other.

Because there is only one pass of combustion products, the heat transfer to water is not particularly good and efficiencies are likely to be down around 60% (or less).

The boiler will be either vertically or horizontally mounted; on the smaller vertical boilers the burner system is likely to be natural draught, whilst the larger horizontal boilers will be fired by fanned draught.

Steel shell – multi-pass fire tube

As the need for higher efficiency boilers grew, so the multi-pass system was developed to allow more convective heat transfer surface area. By installing more tubes and creating a system whereby the flow of the combustion gases passes through a second, third and even fourth pass of the tube configuration.

Figure 6.8 Typical three pass wet back boiler

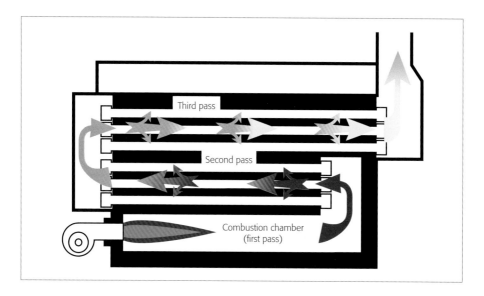

Third pass

Second pass

Combustion chamber
(first pass)

Figure 6.8 details the configuration of a typical three-pass wet back boiler.

The 'wet back' boiler configuration features the water ways surrounding the back end of the combustion chamber and the first row of fire tubes. This arrangement maximises hot boiler surface area and thus heat transfer to water.

Correct water treatment is essential with these boilers to prevent scaling on the water side of the tubes and subsequent 'hotspots' causing possible metal fatigue.

A consideration with this type of boiler when converting from oil firing is the 'back end' temperature associated with the different flame characteristic of a gas flame, compared to that of oil.

The temperature at the back end of the combustion chamber can be 150°C higher with a gas flame causing local overheating at the tube ends of the second pass.

Some boiler manufacturers may specify some course of preparation to reduce the ends of the tube flush with the welds to minimise problems associated with this condition. When converting any boiler, always seek advice from the respective boiler and burner manufacturers – fuel conversion of boilers is reviewed later in this Part.

An alternative system of achieving three passes of the combustion gases is illustrated in Figure 6.9 which features the reverse flow boiler.

The reverse flow boiler features the gas flame firing into the combustion chamber with the POC turning back within the combustion chamber to form the second pass. The third pass is then from the exit of the combustion chamber into the fire tubes.

The benefit with this arrangement is that the POC entering the tubes are sufficiently reduced in temperature to avoid the problems associated with high back end temperatures that can arise with the arrangement featured in Figure 6.8.

Figure 6.9 Principle of reverse flow three pass boiler

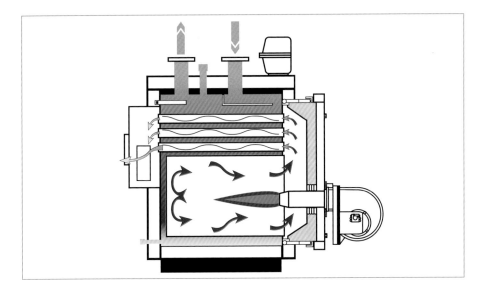

Figure 6.10 Alternative configuration of reverse flow three pass boiler

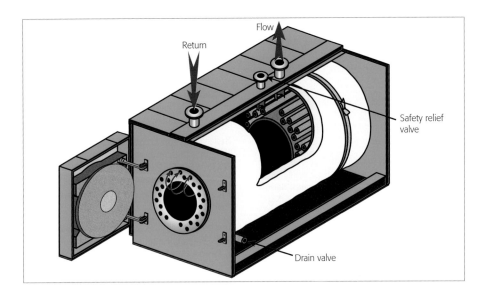

Figure 6.11 Section through typical water heater units

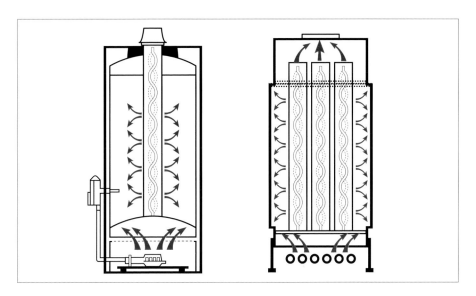

The illustration in Figure 6.9 details the fire tubes located above the combustion chamber, yet another alternative is to locate the tubes all around the periphery of the combustion chamber as illustrated in Figure 6.10.

Types of hot water storage units

Individual storage units

These units generally comprise of a vertical glass lined welded steel cylinder with flue tubes passing through it and an atmospheric burner firing underneath, although some manufacturers do produce forced draught units.

Generally, the multi-tube heaters provide faster heat-up and recovery times to provide rapid response to the system demands.

Figure 6.11 details a schematic view of both single tube and multi-tube heaters.

Each fire tube is usually fitted with a swirl or baffle plate to maximise heat transfer to water.

Cylinders are generally fitted with either removable magnesium sacrificial anodes or an electrical anode protection system to provide protection against metal corrosion in the event of the glass lining becoming damaged.

The removable magnesium anode rods are usually removed from the top of the heater, which means that sufficient height from floor to ceiling will be required to ensure their satisfactory removal. Where room height is restricted, flexible rods are available as optional extra.

Alternatively, and where areas of the country have particularly soft water, the electrical anode system will need to be fitted.

Figure 6.12 Typical direct-fired storage water heater (atmospheric burner)

Figure 6.13 Room-sealed (Type C) storage water heater

Figure 6.12 illustrates a typical direct gas-fired storage water heater with atmospheric burner.

With the wide variation of capacities available, invariably these water heaters can be designed to accommodate specific applications and in many cases can be located close to the point of use, thereby eliminating inefficiencies due to hot water distribution losses.

Available for NG and LPG, the natural draught atmospheric burner system can be fitted with a flame supervision system from that of a simple permanent pilot thermo-electric valve up to fully automatic with spark ignition and flame rectification for applications with interlocks such as exhaust fan and flue damper devices.

Whilst most units are conventionally open-flued Type B (see **Installation requirements – Chimney systems – Classification of flue system** in this Part), some manufacturers offer alternative vertical or horizontal room-sealed systems (Type C), see Figure 6.13.

Modular storage units

If a greater output capacity is required than that of the largest output units, a number of the units can be interconnected as required to give the most flexible and energy efficient arrangement.

Part 4 – Hot water heating systems in this manual reviews the various system options for interconnecting a number of the standard units together.

Alternatively, where the demand priority is for increased volume availability rather than recovery rate, it is possible to interlink a water heater unit (or units) with a storage cylinder.

The storage cylinder will be built to the same specification as the direct-fired heater, but without the obvious requirements of the gas burner and associated controls.

Figure 6.14 Typical rapid heat recovery unit

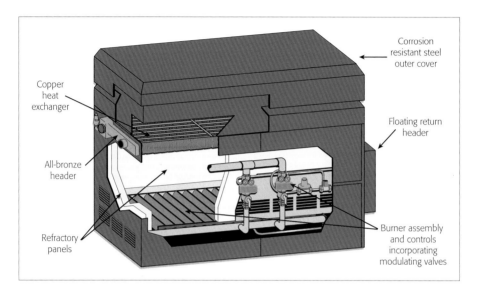

Corrosion resistant steel outer cover

Copper heat exchanger

Floating return header

All-bronze header

Refractory panels

Burner assembly and controls incorporating modulating valves

Rapid heat recovery units

This type of low water content heater comprises of a number of copper or stainless steel finned tubes built-up in numbers into the heat exchanger with cast iron or stainless steel header and manifold.

They are available in a wide range of heat output to water capacities, from about 60kW up to 1MW, with heat recovery of up to 20,000 litres/hr at a 44°C temperature rise (see Figure 6.14).

Again the smaller, most basic models have under-firing atmospheric burners with permanent pilot and thermo-electric valve; as the units increase in output capacity, so the burner system will increase in complexity with fully automatic atmospheric or forced draught burners.

For increased efficiency the burner systems can modulate both gas and air from as low as 20% of full load and during periods of standby the primary air damper can modulate to the closed position to reduce heat loss through the system.

Where required, the heaters can be interconnected with a storage vessel, typically as illustrated in Figure 6.15, to increase volume availability for any instantaneous draw-off.

Figure 6.15 Typical water heater interconnected with storage vessel

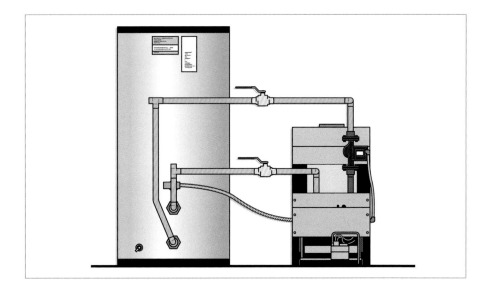

Figure 6.16 Vertical packaged steam boiler

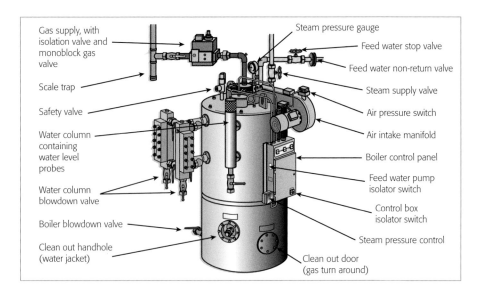

Types of steam raising boilers

Packaged boiler units

These packaged units are generally available from outputs of 96kg/hr up to 960kg/hr and so represent a wide range for the smaller process steam user.

The relatively simple design and robust construction of this boiler makes it ideally suited for the smaller industrial process applications such as laundries, small local bakeries, sterilising processes in the food industry etc (see Figure 6.16).

The application of this type of boiler will normally fall outside the scope of the ACS, in terms of operative competency requirements. However, if the boiler application is that of a large catering establishment, it is likely to fall within the scope of the ACS category COMCAT 2 which covers installation; exchange; disconnect; service; repair; breakdown and commission of pressurised water boilers and steamers.

In addition to any ACS assessment, there will be a requirement for operative training to an ACOP as required by the HSWA.

The top mounted burner fires down into a vertical chamber, which is contained within the pressure vessel. The burner is designed to create a 'spinning' flame for greater turbulence and thus maximise heat transfer to the water.

The flame reverses at the base of the chamber and then passes through the secondary heating space between the pressure vessel and the externally insulated jacket. Vertical fins are welded to the pressure vessel to increase the heat transfer in the secondary pass of the combustion gases.

The burner and all ancillary equipment is pre-assembled by the manufacturers prior to site delivery, hence delivered as a complete assembled package.

With the capacities available, invariably these boilers can be designed for specific applications and in many cases can be located reasonably close to the point of use, thereby eliminating/minimising inefficiencies due to steam distribution losses.

Single and multi-pass steel shell boilers

These types of boilers have been reviewed earlier in this Part for use on LTHW, MTHW and HTHW systems.

Water tube boilers

Water tube boilers tend to be utilised when pressures in excess of approximately 17bar and/or outputs over 18MW and 20,000kg/hr are required, e.g. for power generation. As such the application falls outside the scope of the ACS, in terms of operative competency requirements – specialist training will be required for operatives who work on these systems.

A water tube boiler is essentially a series of drums and headers interconnected by a series of water tubes. Hot combustion gases from the multiple burners pass around vertical banks of water tubes, which are connected to a water drum at the bottom and to a steam manifold drum at the top.

Wet saturated steam is passed through a separator to extract the dry steam, which is then re-circulated through the furnace to raise the temperature of the dry steam to that of superheated steam.

Installation requirements

The following information provided in this Part offers general guidance for the installation of the types of boilers previously described. Specific information relating to the installation of any particular boiler or system will be described in the appropriate manufacturer's instructions.

Particular installation normative reference documents include:

- BS 6644: 2011 'Specification for the installation and maintenance of gas-fired hot water boilers of rated inputs between 70kW (net) and 1.8MW (net) (2nd and 3rd family gases)'

- BS 5440-1: 2008 'Flueing and ventilation for gas appliances of rated input not exceeding 70kW net (1st, 2nd and 3rd family gases) –

  Part 1: 'Specification for installation of gas appliances to chimneys and for maintenance of chimneys'

- IGEM/UP/2 (Edition 2) 'Installation pipework on industrial and commercial premises'

- IGEM/UP/10 (Edition 3) with amendments October 2010 'Installation of flued gas appliances in industrial and commercial premises incorporating specific requirements for appliances fired by bio-fuels'

- EH40 Workplace Exposure Limits: Containing the list of workplace exposure limits for use with the Control of Substances Hazardous to Health Regulations 2002 (COSHH) (as amended) – available from HSE Books – see **Part 12 – References** for contact details.

When designing any hot water heating system there are many areas of consideration. Collaboration will be essential between those involved, such as the gas operative; the occupier; the building owner. There may also be the need to inform local authorities for appropriate planning consent (e.g. for listed buildings), fire authorities, building insurers etc.

Matters to be considered include:

- boiler manufacturer's installation requirements

- building construction (fabric heat losses, suitability and proximity of materials etc.)

- building limitations and its effect on choice of boiler/chimney/ventilation systems

- chimney requirements

- ventilation requirements

- gas availability (type, site capacity and pressure)

- electricity requirements

- site access and liaison with associated trades, etc.

Note: The list should not be regarded as exhaustive.

Competence

Work carried out relative to the installation of boiler heating systems, needs to be undertaken by operatives who are experienced in such work and who hold the relevant certificates of competence, obtained through the ACS and/or aligned N/SVQ's.

Note: For more information regarding commercial ACS categories refer to Part 3 – 'Competence' and Essential Gas Safety – Non-Domestic (Order Ref: ND1, see Part 13 – CORGI*direct* Publications).

Appliance suitability

Any appliance or system must be fit for the purpose for which it is intended. This can be ascertained from a variety of sources:

- CE mark – ensure the appliance carries a CE mark together with the identification number of the notified body responsible for Quality Control surveillance

- manufacturer's data – check that appropriate instructions are available and that the appliance displays a data plate with essential information – maximum input gas pressure, burner gas pressure, kW heat input rating etc.

- the aforementioned data plate should also display;

  - country of import (e.g. GB)

  - type of gas and supply pressure (e.g. G20 = natural gas, G30 = butane, G31 = propane)

  - appliance system category as defined in BS EN 437: 2003 + A1: 2009 'Test gases. Test pressures. Appliance categories' (e.g. I_{2H}, I_{2H3P} etc.).

Air supply/ventilation

To achieve complete combustion it is essential to ensure that the appliance is supplied with an adequate supply of clean fresh air (ventilation), in accordance with the manufacturers instructions.

This will need to take into account the type of flue fitted to the appliance and any other appliances in the same area, along with any cooling requirements for the appliance and the occupants of the workspace in which the appliance is sited.

Note: For information relating to the general ventilation requirements in non-domestic establishments see Essential Gas Safety – Non-Domestic (Order Ref: ND1, see Part 13 – CORGI*direct* Publications).

In addition to manufacturer's instructions, the industry standards referred to for ventilation installations relative to hot water boiler heating systems are as follows:

- BS 6644

- IGEM/UP/10 (Edition 3); and

- BS 5440-2 (for installations below 70kW net)

Gas installation

The GSIUR require gas fitting operatives to ensure that gas installation pipework and fittings are installed safely, with due regard to the location of other services e.g. other pipes; pipe supports; drains; sewers; cables; conduits and electrical control equipment.

Gas operatives will also need to be aware of the limitations of the building structure when installing gas equipment and pipework.

Note: For information relating to the general requirements for gas pipework in non-domestic establishments, see Essential Gas Safety – Non-Domestic – Installation of pipework and fittings (Order Ref: ND1, see Part 13 – CORGI*direct* Publications).

Information in this Part is relevant to the design parameters, materials and methods for the installation of gas pipework and fittings at non-domestic premises. It deals with gas pipework within an industrial or commercial establishment from a meter installation outlet.

The industry standard referred to generally for non-domestic pipework installations is IGEM/UP/2 (Edition 2).

Gas connection

Regulation 26(2) of the GSIUR states –

"No person shall connect a flued domestic gas appliance to the gas supply system except by a permanently fixed rigid pipe."

However, Section 13 of IGEM/UP/2 (Edition 2) states "The use of a flexible connection shall be considered in situations where it is known, or anticipated, that pipework will be subject to vibration, movement, expansion or strain."

It is reasonable to expect that some burner/boiler installations will be subjected to a certain amount of vibration; high resistance shell boilers will often require a gas booster to lift gas pressure to that required by the burner manufacturer, boosters inherently are subject to a certain amount of vibration.

Guidance should be sought from the appliance manufacturer as to the need for gas flexible connection(s).

Where manufacturers stipulate that the final connection to the burner is by means of a suitable gas flexible pipe, the flexible pipe material being employed needs to be stainless steel to:

* BS EN ISO 10380: 2003 'Pipework. Corrugated metal hoses and hose assemblies', Type 1; and

* BS 6501-1: 2004 'Metal hose assemblies. Guidance on the construction and use of corrugated hose assemblies' – minimum acceptable quality 'type B class 1'.

For added protection, it is advisable that the flexible pipe connector incorporates a surface cover. This may be a plastic sleeve, or if there is risk of physical damage, over-braided stainless steel.

Any gas flexible pipe needs to be installed such that the flex is not subjected to stress or torsion.

A union or flanged joint will be required at one end of the flexible pipe to facilitate disconnection of the gas pipe for maintenance of the burner. Installation should ensure that the flexible pipe does not twist when the union is tightened.

The route of the gas pipe, connecting to the burner will need to be located such that it does not impede withdrawal of the burner for maintenance purposes.

In any event, a 90° operable valve will need to be installed immediately upstream of the flexible pipe assembly.

Due consideration should be given to the available gas supply pressure in the pipe system feeding the boiler.

On some industrial and commercial sites it may be possible for the metered supply to have elevated gas pressure, or on large sites, such wide variation on load characteristics as to create significant variation in the gas line pressure.

Where either of these situations exist, it will be necessary to check with the appliance manufacturers to establish if there is a need for fitting an additional gas regulator upstream of the appliance control system.

In certain cases, particularly where the supply pressure to the appliance is 75mbar or more, there will need to be adequate automatic means for preventing the appliance and associated pipework from being subjected to this higher pressure e.g. in the event of regulator diaphragm failure. This would usually be by means of a regulator with an integral overpressure slam-shut valve incorporated.

Electrical connections

Non-domestic properties normally utilise three-phase electrical systems due to the high current demands of the electrical equipment installed and used.

When gas operatives are required to work on gas appliance/equipment, which utilise either three-phase or single-phase electricity, it is important that those gas operatives are competent and have the relevant knowledge and experience to work safely on those systems.

They need to be able to carry out a thorough risk assessment of any potential hazards that may be involved. For further information on risk assessments see Essential Gas Safety – Non-Domestic – Part 3 – 'Risk assessments' for further guidance.

Important: The Electricity at Work Regulations 1989 (EWR) apply to any operative carrying out any type of electrical work. It is a requirement of the EWR that the person is competent to undertake the work.

Where any doubt exists with regards to competency or lack of knowledge/experience to work on those electrical systems, or where a risk assessment identifies that the hazard would be unacceptable, no work should be undertaken.

Working on three-phase equipment, supplies and four wire supplies

In the UK the 400V, 4-wire three-phase and neutral (TPN) method is widely used for the distribution of supplies within commercial and small industrial installations.

For example, on large wet central heating systems the pump may operate at 400V from a three-phase supply while other components such as thermostats and time switches may be 230V single-phase or sometimes even less, for example 24V, which is classed as Extra Low Voltage (ELV).

On commercial warm air heaters, the fans (combustion and/or circulating) may be three-phase, while other components may be single-phase.

It must be remembered that each of these phases is carrying 230V. A balanced three-phase circuit does not require a neutral for it to be able to operate. This is because there is always a return path for the current through one of the other two phases.

There are a number of signs to look for when trying to identify the voltage on non-domestic gas installations, two most basic signs are:

1. Appliance data plate – one of the simplest methods of determining the voltage to a non-domestic gas appliance is by checking the information on the appliance data plate, assuming of course a plate is present.

    The information displayed will typically consist of:

    - supply voltage – 400V

    - supply frequency – 50Hz

    - no of phases – 3

    - no of wires – 4

    - power consumption – 4.0kW

    - fuse rating – 400V, 25A HBC

    - fuse rating – 230V, 5A

2. Installation wiring – all installation wiring falls within the scope of BS 7671: 2008 (Incorporating Amendment No.1 2011) 'Requirements for electrical installations. IET Wiring Regulations Seventeenth Edition' and therefore requires completion of electrical certification when installed.

    The existence of a three-phase supply can often be confirmed by the wiring colour code. This will need to be tested to ensure it is correctly sized etc. before undertaking any electrical installation work.

Remember, all electrical installations should be in accordance with BS 7671 and the supplied manufacturer's wiring guides, especially for the correct connection of supplies. Of equal importance is the correct routing of cables, including suitable supports where required and the correct securing of cables to individual pieces of equipment.

Each appliance and its associated controls needs to have its own means of electrical isolation, which should be situated in close proximity to the equipment it serves and have suitable means for the application of securing devices, i.e. hasps or padlocks for ensuring that the appliance cannot be re-energised whilst being worked on by an operative (this should be highlighted as part of a risk assessment).

Additional safety controls

There can be occasions when additional safety controls other than those integral to the boiler are thought necessary.

The function of these safety controls will be to ensure the boiler is automatically isolated in the event of a potentially dangerous condition arising; the type of situation that could occur at roof top boiler installations resulting from severe weather conditions for example.

Where it is felt necessary to install any such additional safety control then the following are required:

* to shut off the gas supply. The system needs to be reset manually; and

* a warning light, or other such signal, will need to be operated to show that a fault condition has arisen drawing attention to the situation so that appropriate action can be considered. The signal to remain activated until such time as the shut down condition has been rectified and the control reset.

Valve proving systems

BS EN 676: 2003 + A2: 2008 (Incorporating Corrigendum December 2008) 'Automatic forced draught burners for gaseous fuels' is primarily intended for automatic forced draught gas burners having a combustion air fan, operated with gaseous fuels and intended to be marketed as a complete assembly. It is therefore a document intended for burner manufacturers to comply with.

However, a point for operatives to note in the standard is that all such burners are required to be fitted with two automatic safety shut-off valves (SSOVs) on the main gas feed of the appropriate classification; class 'A' or 'B' as defined in BS EN 676, Table 1.

In addition, where the burner input is rated at more than 1200kW, the required class 'A' valves will need to incorporate the operation of an automatic valve proving system prior to any attempt at ignition.

Note: For information relating to the operation of SSOVs and valve proving systems on boilers and plant in non-domestic establishments, see Essential Gas Safety – Non-Domestic (Order Ref: ND1, see Part 13 – CORGI*direct* Publications) and IGE/UP/12 Application of burners and controls to gas fired process plant.

It should be noted that the former British Standard for automatic burners above 60kW (gross) – which has been partially replaced by BS EN 676 – was BS 5885-1: 1988 'Automatic gas burners. Specification for burners with input rating 60kW and above' only required valve proving systems to be fitted on burners rated above 3MW; between 1MW and 3MW a system of valve position indication was required.

Therefore, when operatives encounter some of the older boiler/burner arrangements (pre 1996) rated between 1.2MW and 3MW they will not usually comply with the current BS EN 676.

Equipment in the water circuit

If not supplied by the manufacturer as part of the boiler assembly, there needs to be facilities provided for safely isolating the water supply to each boiler or bank of boiler modules in the case of modular boiler systems.

If three way valves are to be used to isolate the water supply, the flow through the boiler will need to be open to atmosphere through the third port.

Where multiple boilers are installed, other than modular boilers, each boiler connected to the common flow and return system headers will need a facility for isolating it from the system. In the case of modular boilers this facility can be on each bank of modules.

Again, if not supplied by the manufacturer as part of the boiler assembly, each boiler will need to be fitted with a safety valve conforming to BS EN ISO 4126-1: 2004 'Safety devices for protection against excessive pressure. Safety valves'.

In the case of modular boiler systems a common valve can be fitted on each bank of boilers. This safety valve will need to be set to lift at a pressure, which is no greater than the maximum allowable pressure for any component in the system.

If the safety valve is not fitted directly to the boiler, then the bore of the pipe connecting it to the boiler will need to be of at least the same diameter of the bore of the safety valve. No other valve or tap can be fitted between the boiler and the safety valve, so as to form any kind of restriction.

The discharge pipe from the safety valve is required to be self-draining and terminate in a visible position. However, due consideration will need to be given to ensure that any discharge does not represent a hazard to personnel or affect the plant, e.g. cause corrosion.

For details on correct sizing of the safety valve and discharge pipe, operatives will need to refer to BS 6644.

Open vented systems

The cold feed pipe will need to run internally and be taken directly from the feed and expansion cistern to the boiler and the feed cannot supply water elsewhere for any other purpose. The cold feed to multiple or modular boilers will need to be fitted with a lockable isolating valve.

Where the cold feed is routed in unheated areas or where condensation can form, the pipe will need to be suitably insulated.

Open vent pipes will need to take the shortest practicable route and be installed in a manner which does not induce any restriction or obstruction, e.g. no isolation valve fitted, which could prevent the safe venting of the boiler during operation.

Any valve fitted between the boiler and the open vent to facilitate maintenance will need to incorporate a three way valve such that when closed to the vent pipe, the boiler is open to the outside atmosphere through the third port; the valve needs to include means of indicating the position of the open port.

For details on correct sizing of the vent pipes, operatives will need to refer to the BS 6644.

Sealed systems

For details concerning gas cushioned systems, operatives will need to refer to BS 6880–2: 1988 'Code of practice for low temperature hot water heating systems of output greater than 45kW. Selection of equipment'.

Any diaphragm expansion vessel incorporated in the sealed system will need to comply with BS EN 13831: 2007 'Closed expansion vessels with built-in diaphragm for installation in water' (replaces the previous BS 4814 standard) – and installed in accordance with manufacturer's instructions.

The connecting pipe between the vessel and the system cannot be fitted with any valve unless that valve can be locked in the open position.

There will need to be a suitable provision for replacing lost water from the hot water system and details of such systems are provided in BS 6644.

The cold water feed will need to incorporate:

*   a verifiable backflow prevention device to fluid category 4; or

*   a device no less effective; and

*   non-return valve and an isolating valve

There will need to be an automatic air venting device fitted between the isolating valve and the non-return valve with the isolating valve located between the air venting device and the system pipework.

The following controls will need to be installed:

*   low pressure water cut-off device to shut down the boiler (interlocked) in the event of pressure loss due to significant water leak

*   water pressure gauge sited so it can be easily read and replaced without draining the system

*   temperature gauge sited so it can be easily read (in degrees celsius) and replaced without draining the system

*   drain valve fitted with a removable key

*   if a high water pressure cut-off device is fitted it will need to shut down the boiler (interlocked) in the event of over pressure.

Chimney systems

Changes to industry standards – note new definitions

BS EN 1443: 2003 'Chimneys. General requirements' provide some new definitions which alter common terms we have been used to within the gas industry, when we talk about flues and chimneys.

The following definitions are used within this manual:

Chimney: This is a structure consisting of a wall or walls enclosing the flue or flues i.e. includes room-sealed flues

Chimney component: This is any part of a chimney.

Flue liner: This is the wall of a chimney consisting of components, the surface of which is in contact with the POC. This is not just a flexible flue liner but also any suitable material to convey these products.

Flue: This is the passage or space for conveying POC to the outside atmosphere.

Classification of flue systems

Appliances and their flues are classified by a 'European Committee for Standardisation' document entitled PD CEN/TR 1749: 'European scheme for the classification of gas appliances according to the method of evacuation of the products of combustion (types)'.

The purpose of this document is to harmonise, across Europe, the classification of appliances burning combustible gases.

PD CEN/TR 1749 separates gas appliances into three types:

*   Type A – flueless

*   Type B – open-flued (including systems formerly known as 'closed' flue)

*   Type C – room-sealed.

For further details on the correct flue classification for general appliance types, see Essential Gas Safety – Non-Domestic (Order Ref: ND1, see Part 13 – CORGI*direct* Publications).

Chimney installation

Any chimney or flue pipe system will need to be installed in accordance with manufacturer's instructions to ensure safe and complete evacuation of the POC to the atmosphere. Chimney components and their materials of construction will need to be in accordance with the appropriate British Standards.

Note: For details on the general appliance chimney installation, see Essential Gas Safety – Non-Domestic (Order Ref: ND1, see Part 13 – CORGI*direct* Publications).

Single boiler chimney installation – Type B

Unless otherwise stated in manufacturer's instructions, it will be necessary to install the chimney/flue pipe so that it is supported properly throughout its length and independently of the boiler.

The manufacturer's chimney adaptor piece should always be used to ensure correct connection to the flue spigot without reducing the cross sectional area (csa) and that a complete seal is obtained. Consider also, that provision will need to be made so that the chimney/flue pipe can be disconnected from the boiler to allow inspection and servicing.

The flue terminal will need to be positioned so as to ensure that no POC are allowed to re-enter the building or re-circulate into the combustion air intake. If the terminal is located less than 2m above ground level (e.g. wall termination), then it needs to be fitted with a suitable guard to prevent possible blockage or injury to persons.

For boilers rated below 70kW net, the flue termination will need to conform to the requirements of BS 5440-1.

Note: It is generally recommended that consideration be given to the fitting of a suitable terminal where the flue terminates in free atmosphere. However, where the flue diameter is less than 170mm an approved terminal will be required.

The route of the chimney for natural draught boilers will need to be such that there are no downward sections or downward incline throughout its length.

Where the route of the chimney is such that horizontal runs are unavoidable, the appliance manufacturers should be consulted regarding maximum horizontal lengths.

With natural draught appliances it is usually possible to purchase an exhaust fan from the manufacturer for the particular boiler to be installed, which will assist in removing the POC. Any such fan will need to be installed in accordance with the manufacturer's instruction, particularly with respect to ensuring that the gas burner will shut down if the fan fails (interlocked).

Prevention of condensation within the flue is an important design factor, manufacturers will usually recommend the use of an approved double wall flue pipe to minimise the likelihood of this occurring.

Where condensation in the flue is unavoidable or part of the design (i.e. condensing boilers), appropriate provision will need to be made for the condensate to flow freely to a disposal point. Additionally, the chimney/flue pipe/flue liner, as appropriate will need to be manufacturered from material impervious to condensation and corrosion resistant.

Current Standards/Regulations require that a data plate is displayed on existing chimney systems detailing the various performance characteristics of that chimney and its flue.

Where a boiler is to be connected to such an existing chimney/flue pipe, the gas operative will need to verify with the boiler manufacturers that the information displayed on the data plate is appropriate for the connection of the proposed boiler(s).

Multiple boiler chimney installation – Type B

Where two or more boilers are connected to a common natural draught flue, the boilers need to be installed in the same room and be fitted with the same type of burner system, i.e. forced draught burners cannot be connected to a common flue system shared with atmospheric burners. It is generally recommended that the maximum number of modular boilers connecting to a common flue is 6.

Where one boiler is likely to fire for longer periods than the others it will need to be located nearest to the main vertical flue.

For energy efficient operation of the modular system some manufacturers now incorporate an automatic damper in the flue exit on the boiler. This damper closes when the burner is off to minimise standby losses from the boiler, fully opens when the burner is on high-fire and throttles back to restrict the pull on the flue and hence the amount of air drawn into the combustion chamber when on low-fire. The damper mechanism will need to be fully interlocked with the burner so that the burner cannot fire with the damper restricted or closed.

Any manual dampers fitted in the vertical sections of flues will need to be set and locked in position during the commissioning process.

Where a flue pipe enters a common chimney system, the method of connection will need to ensure that no interaction of combustion products between appliances can take place. Ideally the connection between any individual flue and the common header needs to be a smooth bend or a sloping connection at 135°.

There will need to be at least 500mm of vertical secondary flue pipe from the base of the draught diverter and the connection to the header.

For other details regarding the installation of common chimney/flue pipe systems for modular boilers, see Essential Gas Safety – Non-Domestic (Order Ref: ND1, see **Part 13 – CORGI**_direct_ **Publications**).

Single boiler chimney installation – Type C

Any room-sealed flue system will need to be assembled and installed with manufacturer's approved components.

When using concentric flue termination, the complete system will need to be installed with manufacturer's approved components and in accordance with the manufacturer's instruction.

This type of system is regarded as an integral part of the appliance and forms the basis on which the appliance was tested to achieve the CE mark. To deviate from this would therefore contravene the GSIUR.

The flue terminal will need to be positioned so as to ensure that no POC are allowed to re-enter the building or re-circulate into the combustion air intake. If the terminal is located less than 2m above ground level, then it needs to be fitted with a suitable guard to prevent possible blockage or injury to persons.

Fuel conversion

There are many reasons for considering the possible conversion of an existing boiler from coal or oil firing to gas firing, which can include:

• fuel costs – whilst the commodity price for gas is usually at a premium and will vary from month to month based on spot market prices, the excellent controllability of boilers fired by gas can significantly improve operating efficiencies and can lead to cost savings

- energy conservation – higher thermal efficiencies can be obtained with gas firing

- attendance – solid fuel fired boilers and possibly heavy fuel oil fired boilers generally need some form of manual attendance on a regular basis so there is often a labour saving associated with a change to gas firing

- fuel supply – the manual ordering process and supervision of delivery is eliminated

  - gas is supplied direct to the point of use

  - no on site storage required for NG

  - no waste or ash removal necessary

- maintenance requirements – are generally less for gas firing, an annual service visit is often all that is required for gas-fired boilers

- pollution – if the burning process is more controllable with gas firing it follows that there will be less Carbon Dioxide (CO_2) and other combustion emissions to pollute the atmosphere.

Dependent on the age and condition of the existing boiler(s), greater efficiencies and therefore economies might be achieved by replacing the boilers with modern gas-fired modular and/or condensing boilers.

However, often all that is required is a simple burner change, particularly if the original fuel is 35 second oil.

Considerations for conversion are many and varied but should include the following:

- availability of a suitable gas supply at the correct pressure and delivery period

- current fuel usage in terms of type, grade and consumption – this will assist in determining the correct size gas supply

- obtain information on any future expansion planned for the premises

- inspect past insurance reports – these will give some guidance as to the general condition of the boiler. Also the insurer will need to be advised of any proposed conversion

- obtain boiler manufacture's recommendations for conversion. Most boiler manufacturers will have trialled several burners to obtain the best match for the boiler and can produce a list of preferred manufacturers and recommended burners; likewise for burner manufacturers.

  In certain cases where the boiler was originally specifically designed for oil firing, the combustion chamber may not be large enough for gas firing. It might be necessary to down-rate the boiler when converted to gas firing due to the possibility of flame impingement

- check boiler manufacturer's recommendations for treatment of first pass tube ends on steel shell boilers

- check existing chimney system for suitability; height, diameter and material, also any signs of leakage of the POC (spillage)

- if a manual flue damper is fitted, can it be removed or locked in the open position. Any automatic damper will need to be interlocked with the proposed gas burner control system

- if there is an existing exhaust fan, ascertain if possible, its duty and ensure it is suitable and can be interlocked with the proposed gas burner control system

- check that existing ventilation facilities will be adequate for the gas burner

- note the condition of any existing insulation on the boiler and associated pipework, observe particularly if it is made of asbestos.

Note: This list should not be regarded as exhaustive.

Commissioning

Manufacturer's instructions will provide specific information for commissioning any particular appliance and its associated equipment. In addition, IGEM's publication IGE/UP/4 (Edition 2) 'Commissioning of gas-fired plant on industrial and commercial premises' offers further generic guidance for the commissioning process.

Correct commissioning of an appliance in accordance with the manufacturer's instructions is as critical to a safe and efficient installation, as its initial installation.

Therefore, where the installation can not be commissioned immediately after installation, it should be isolated from the gas supply and suitable notification attached (this fact also needs to be documented on any appropriate paperwork) to advise the reader that the installation is un-commissioned.

The following is provided for information purposes only and should not be regarded as substitute for the source documents.

Planning

Before travelling to site, a certain amount of preparation will be necessary, including:

- ensure all relevant manufacturer's information is available

- all relevant drawings and system plans are available

- all gas tightness test/purge certificates have been completed by the upstream pipe gas operative

- all electrical test certificates have been completed by the competent electrical operative

- any pressurised flue system has been correctly tested by the manufacturer's approved operative and documented

- risk analysis has been completed

- COSHH statements prepared where relevant

- any hot work or permit to work has been granted

- all necessary tools are available and suitable for the purpose and are of sound quality. Any electronic instruments need to be correctly calibrated and certificated accordingly

- a written plan of required work is completed (usually provided in manufacturer's instructions).

Inspection

Once on site, a complete inspection of the installation will need to be carried out to ensure that:

- all gas and electrical supplies are suitably sized, correctly located and supported and that the installation has been installed in accordance with the manufacturer's instructions and relevant standards

- assembly is complete and all components are fit for the intended purpose

- the appliance has been correctly positioned; it is level as required by manufacturer's instructions and is stable with adequate clearance for maintenance and from combustible materials

- the chimney/flue pipe is correctly installed

- adequate ventilation is provided

- gas and electrical supplies have been isolated

- the wet system has been correctly filled, free of air and pressurised

- all necessary safety interlocks are correctly installed and are appropriate, e.g. automatic isolation valves (AIVs), if fitted in the incoming gas line to the boiler room are suitable for the application (see comments regarding the use of AIVs in **Part 5 – Boiler locations**).

Activation – dry run

Subject to satisfactory inspection, the next phase of the commissioning process is to carry out preliminary physical checks with the fuel isolated:

- a suitable gas tightness test is undertaken on the appliance connector pipe.

Note: For further detail on the correct procedures for testing gas pipework in non-domestic establishments, see Essential Gas Safety – Non-Domestic (Order Ref: ND1, see Part 13 – CORGI*direct* Publications).

The industry standard referred to generally for non-domestic pipework testing and purging is the IGE/UP/1 (Edition 2) 'Strength testing, tightness testing and direct purging of industrial and commercial gas installations' or alternatively if within its scope, IGE/UP/1A (Edition 2) 'Strength testing, tightness testing and direct purging of small, low pressure industrial and commercial natural gas installations'.

- check that all manual isolating valves and safety shut off valves are closed and are leak tight (e.g. no let-by)

- any control interlocking device is set to its provisional operating level, considered safe for commissioning (e.g. regulator, process controls and interlocks)

- with auxiliary energy supplies available, all power equipment and interlocks are checked for operation, e.g. where automatic control units are employed, the controller is checked at each stage of its light-up sequence for correct operation and response of the control unit including automatic valve proving system where fitted

- check that the system of ignition is adequate, e.g. strength and location of any spark generation is correct to ensure smooth and reliable lighting of start gas flame

- flame safeguard systems correctly go to shutdown condition in the absence of a flame, e.g. lockout within the specified time period. If a simulated flame is present the system detects the presence of the flame correctly.

Activation – fuel run

Once all checks have been made with the fuel isolated to the satisfaction of the commissioning operative, the gas may be turned on and the following further checks initiated:

- the appliance is correctly purged of air (pipework connector and combustion chamber)

- allow the burner system to proceed through to start gas ignition or pilot stage, ensure that start gas flame or pilot is established and set at the correct pressure before proceeding to main gas

- confirm by suitable means of leak detection that all start gas or pilot gas pipework is gas tight

- ensure that loss of start gas/pilot flame results in the correct shutdown of the control system. Check thermo-electric devices for the correct 'drop-out' time as described in manufacturers information

- ensure that, with the burner shutdown, the safety shut off valve(s) remain gas tight

- with main gas available and controls restricted to a nominal main flame ignition rate, allow the burner to proceed to main gas ignition. Ensure that the main gas flame is established and set at the correct pressure

- confirm by suitable means of leak detection that all main gas burner pipework is gas tight

- ensure that loss of gas flame results in the correct shutdown of the control system

- ensure that, with the burner shutdown, the safety shut off valve(s) remain gas tight.

Operational checks

Once all checks have been made to ensure safe and reliable ignition to main gas burner stage, to the satisfaction of the commissioning operative, the following further operational checks can be initiated:

- the heating plant is allowed to run up to normal operational temperatures to ensure the unit remains satisfactory

- operational controls can now be checked for satisfactory operation, e.g. room thermostats

- combustion gases are effectively dispersed and ventilation is adequate

- combustion checks are carried out

- undertake spillage tests where appropriate

- ensure that the requirements of the GSIUR 26(9) are met, i.e. –

  - the effectiveness of any flue;

  - the supply of combustion air;

  - its operating pressure or heat input or, where necessary, both; and

  - its operation so as to ensure its safe functioning

When undertaking combustion analysis, the correct instrumentation will need to be used ensuring it is suitable for the application and that correct levels of accuracy are achieved.

Note: For further detail on the correct procedures for combustion analysis and appliance efficiency testing, see Essential Gas Safety – Non-Domestic (Order Ref: ND1, see Part 13 – CORGI*direct* Publications).

The industry standard referred to generally for using portable electronic combustion gas analysers is BS 7967.

The standard currently comprises of 5 Parts. Parts 1-4 deal with domestic installations and Part 5 being applicable to non-domestic installations:

Part 5   Carbon monoxide in dwellings and other premises and the combustion performance of gas-fired appliances. Guide for using electronic portable combustion gas analysers in non-domestic premises for the measurement of carbon monoxide and carbon dioxide levels and the determination of combustion performance

Completion

Once all aforementioned checks have been made to the complete satisfaction of the commissioning operative, the following will be required:

- all users are instructed in the correct operation of the boiler and/or system and its user controls, light-up and shutdown sequence

- manufacturer's 'user' instructions are left for the responsible person. The 'responsible person' is defined by the GSIUR as the 'owner or occupier'. However, on many large commercial or industrial sites this may not be relevant, it may be more appropriate to identify the works engineer or manager for this purpose

- the responsible person should also be advised concerning matters such as actions to be taken in the event of fault or emergency conditions and advice concerning regular maintenance of the units

- a suitable report will need to be completed and left with the responsible person. This report needs to detail the final setting parameters of the heater/system, including;

    - gas user/site details

    - plant/boiler details, e.g. make, model and serial number

    - fuel supply details, e.g. type and supply pressure

    - appliance/system operating set levels

    - combustion and emission data

    - electrical data, e.g. nominal supply, overload settings, fuse ratings.

Note: CORGI*direct* produce a suitable form for recording the commissioning process – 'Plant Commissioning/Servicing Record' (Order Ref: CP15, see Part 13 – CORGI*direct* Publications).

Handover

Upon satisfactory completion of all the aforementioned commissioning actions, the commissioning report, together with any manufacturer's installation and maintenance instructions are handed over to the designated responsible person for the site premises.

It should then be made clear to all concerned that the commissioning process has been completed and responsibility for the plant is passed on to the appropriate personnel.

Overhead radiant heating – 7

7 – Overhead radiant heating

Principles of radiant heating

Radiation does not rely on the presence of an intervening medium to transfer heat. It does so by the transference of heat in electromagnetic waveform similar to light. The particular method of heat generated by a gas-fired heater unit uses infrared rays.

Gas-fired infrared heaters radiate on 0.8 to 500 micrometres wavelengths.

Radiant infrared energy acts like light in that it radiates in all directions from its source and travels in straight lines. So the infrared rays can be directed downward to target and heat objects such as the floor, walls, machines and people.

When these objects are heated, they in turn release heat to the atmosphere, reflecting the radiated heat as low power radiators.

When the radiated heat is emitted from a source, part of the energy is reflected by the receiving body and part is absorbed. If two bodies are placed in a confined space, with one body hotter than the other, there will be a continuous interchange of energy between them. The hotter body will radiate more than it absorbs and the cooler body will do the opposite.

Ultimately, if the process were to continue without external influences, the two bodies would reach a steady state whereby the rates of radiation and absorption become the same.

Convection also takes place, transferring heat from the warm floor or bodies to the cool air lying at low level. As cool air flows across the surface of the bodies, it is heated and rises and is displaced by cooler air.

This continuous cycle gradually raises the air temperature within the building and given proper control, can maintain acceptable comfort levels.

In this way, all useable energy is absorbed, such that people occupying the area feel comfortably warm, even though the actual air temperature may be relatively cool.

People are warmed from multiple sources, not only are they blanketed by infrared radiation from the heaters, they benefit as a result of secondary radiation from below and warm air rising as a result of convection from the heated bodies.

The effect is similar to that of standing in sunshine in the open atmosphere when air temperature is relatively cold.

Radiant heating can therefore offer significant energy savings compared with some other forms of heating systems such as centralised boiler systems, or even some warm air distribution systems.

This is due to the design ability to target heat where it is needed, particularly in large working areas or where there are high levels of air volume changes per hour.

Radiant heating can be used as an effective method of spot heating, for example in a large warehouse area where staff operating in the warehouse are localised in one particular zone within the warehouse.

Alternatively, by building a series of heater units to cover the whole floor area, it may be used as a total heating scheme.

It can be seen from Figure 7.1, by effective design, an area may be heated for the general comfort of the occupants whilst maintaining a cooler ambient temperature than would otherwise be maintained using alternative systems of heating, which heat the total volume of the building, including the roof space which is not occupied (e.g. warm air distribution).

Figure 7.1 Radiant heat pattern

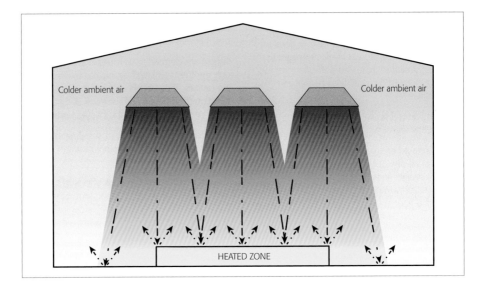

Colder ambient air

Colder ambient air

HEATED ZONE

Although temperature control in a room can be achieved with conventional air thermostats, position of the thermostat becomes very important.

It is often a requirement for radiant systems, to have room temperature controlled by means of a 'black bulb' type thermostat, rather than conventional air thermostats due to the cooler ambient temperature.

Air thermostats would tend to keep the heater units running for longer periods than may be required for general comfort of the occupants if they are not sited correctly.

Radiant systems are widely used for heating areas such as sports halls, warehouses, garden centres/greenhouses, aircraft hangars and factories etc.

The general benefits of radiant heating can be summarised as follows:

- the heat can be targeted to particular zones reducing the level of heat input required to give comfort conditions

- only bodies in the path of the infrared rays are heated, the air is only heated by convection/radiation when reflected from the surface of a heated body

- heat losses through the fabric of the building are minimised, particularly through walls and roof structure above the heated zone

- the heaters rapidly achieve normal running temperatures, thereby reducing overall warm-up periods

- the heaters are usually suspended from the steelwork or fabric of the building at high level and do not occupy valuable floor space

- multiple heater unit systems will invariably mean that in the event of one heater unit breaking down, the rest of the heating system is not compromised.

Figure 7.2 Typical radiant plaque heater

Typical radiant heating systems

Radiant heating systems can generally be categorised into two basic types, namely:

* luminous – high intensity, e.g. ceramic plaques, cone heaters (or brooders, as they may be called); or

* non-luminous (black heat) – low intensity, e.g. radiant tubes, air tube heaters and catalytic panels.

Luminous – high Intensity

Luminous heaters emit rays of short infrared wavelength, approximately 2.5 micrometres.

These units generally utilise a ceramic heat emitter housed in a metal frame with stainless steel or aluminium deflector. An injected gas/air mix is supplied to a burner, which fires onto a ceramic panel, or alternatively, supplied direct to flame ports in the porous ceramic plaque surface.

The heaters themselves are flueless, so the POC rise off the heater panel and are dispersed into the premises.

When the gas is ignited, surface temperatures are generally between 800°C to 1000°C. With such temperatures they are ideally suited for spot heating in large or well ventilated areas.

They are generally available with fuel options of NG or LPG.

Radiant Plaque

Plaques may be suspended horizontally to direct the heat straight down, or mounted on wall brackets to deflect the heat at a predetermined angle (see Figure 7.2) – mounting height will directly affect the radiant intensity.

Typically, radiant plaque heaters can weigh between 6kgs and 30kgs and their input range can be from 3kW up to 30kW.

Figure 7.3 Cone heater

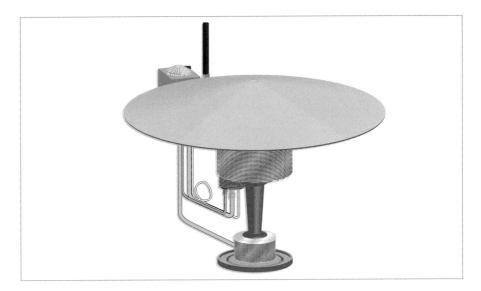

A variety of burner control methods are available:

- lever, cock and chain – the most basic of any design, incorporates a manually lit permanent pilot with thermo-electric flame safeguard, main gas is operated by a manual valve which will be linked via a chain to normal operating level

- semi-automatic – incorporates a permanent pilot and main gas controlled through a thermo-electric multi-functional valve. Gas throughput to the main burner will be electronically controlled via a solenoid valve in the multi-functional valve. With this facility, it will be possible to incorporate time-switch and thermostatic control systems

- fully automatic – incorporates spark ignition of a pilot flame with flame supervision using flame rectification prior to opening of the main gas solenoid valve. Smaller heater units may ignite straight to main gas without the interim stage of lighting the pilot.

Cone heaters (brooders)

Heaters are suspended horizontally to direct the heat straight down.

Typically their input range may be from 3kW up to 20kW and mounting height will directly affect the radiant intensity.

Although this type of heater is often used as a means of spot heating in a workshop or factory area, it is also widely used in the rearing of poultry (see Figure 7.3).

A single unit (nominally up to 10kW) can be used in raising as many as 1500 chicks. In this application, the mounting height of the heater unit will be lower than conventional space heating.

The following burner control methods are available:

- semi-automatic – incorporates a permanent pilot and main gas controlled through a thermo-electric multi functional valve. Gas throughput to the main burner will be electronically controlled via a solenoid valve in the multi-functional valve. With this facility, it will be possible to incorporate time-switch and thermostatic control systems

- fully automatic – incorporates spark ignition of a pilot flame with flame supervision using flame rectification prior to opening of the main gas solenoid valve. Smaller heater units may ignite straight to main gas without the interim stage of lighting the pilot.

Non-luminous – low Intensity

Non-luminous heaters emit rays of a longer infrared wavelength, approximately 4 micrometres.

These units generally utilise a gas burner firing into a mild steel tube and utilise a stainless steel or aluminium deflector. The gas is combusted using a small combustion air fan either at the burner end of the tube, or at the exhaust end dependent on the manufacturer's configuration.

The heaters are generally flued, but for spot heating in large or well ventilated areas, they may be installed without a flue system, so the POC rise off the exhaust and are dispersed into the premises.

When the gas is ignited, tube surface temperatures are generally between 300°C to 450°C at the burner end reducing to 150°C at the exhaust on some models.

Types of radiant tube system include:

- compact 'packaged'
- 'U' tube
- linear (single and multiple burner)
- continuous (CoRayVac, NorRayVac).

Other types of Low intensity radiant heating systems include:

- air heated large bore tube
- catalytic Panels.

Compact 'packaged'

The compact 'packaged' unit represents a relatively new concept in radiant heating technology. It is a unit heater comparable with radiant plaque units in that it requires no on-site assembly work. It can be easily installed in much the same way as any plaque or cone heater, keeping on site handling and installation to a minimum (see Figure 7.4).

These compact units may be suspended similarly to radiant plaque units i.e. horizontally to direct the heat straight down or mounted on wall brackets to deflect the heat at a predetermined angle.

A burner fires into the pre assembled compact tube from above. The tube itself is constructed of stainless and aluminised high temperature steel for longer life. Radiated heat is deflected down by the reflector panel encased above the tube.

The units are available in 12kW, 20kW and 40kW options for firing on NG or LPG and weigh between approximately 90kgs and 150kgs.

They may be installed in a variety of flue configurations:

- flueless
- open-flued
- room-sealed; or
- central 'herringbone' manifold.

If they are to be installed in dusty atmospheres, it is possible to connect a fresh air intake duct to provide combustion air direct from outside atmosphere.

Figure 7.4 Compact radiant tube

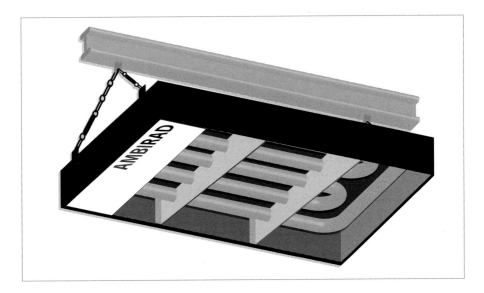

Noise levels are in the region of 48dB, which can further be reduced to approximately 42dB(9A) if fitted with the aforementioned fresh air duct.

If the units are to be installed in such applications as sports halls (particularly badminton courts), they are available with protective guards and decorative grilles.

These units are fully automatic with electronic ignition and flame supervision.

'U' tube

'U' tube units are longer than the compact units, typically between 5m and 7m and weigh between 40kg and 90kg, dependent on heat input rating, which will generally be between 9kW and 50kW for firing on NG or LPG (see Figure 7.5).

They comprise of an aluminised steel or stainless steel tube typically 75mm in diameter assembled with a prefabricated 'U' section at one end of the heater and burner/exhaust at the other. The burner and fan configuration will vary dependent on the manufacturer's individual design.

Most units utilise an induced draught system, whereby an atmospheric 'cup' type burner fires into one end of the tube assembly, with an exhaust fan fitted at the end of the return tube. However, some manufacturers incorporate a forced draught system whereby combustion air is provided by the fan at the burner end.

Radiated heat is deflected down by the stainless steel or aluminium reflector mounted above the tube assembly.

Figure 7.5 'U' radiant heater

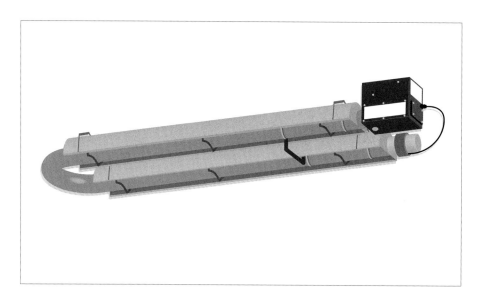

Similar to the compact unit, radiant 'U' tube heaters may be installed in a variety of flue configurations:

- flueless

- open-flued

- room-sealed; or

- central 'herringbone' manifold.

If they are to be fitted in dusty atmospheres, it is possible to connect a fresh air intake duct to provide combustion air from the outside atmosphere.

Radiant 'U' tubes are ideally suited for spot heat applications, a single tube offers floor coverage between 20m² and 400m² based on the aforementioned range of kW rating and manufacturer's specified mounting heights. Multiple units offer the ability to provide heating for large areas up to complete space heating schemes, e.g. for factories or garage workshops.

Linear tube (single burner)

Linear tube units can be longer than the 'U' tube units, typically between 6m and 15m, but the weight will be similar, i.e. between 40kg and 90kg, dependent on heat input rating, which will generally be between 15kW and 50kW for firing on NG or LPG (see Figure 7.6).

They comprise of an aluminised steel or stainless steel tube typically 75mm diameter and most units are assembled with a burner firing into one end of the tube and induced air/exhaust fan at the other.

Radiated heat is deflected down by the stainless steel or aluminium reflector mounted above the tube assembly.

Because of the narrow width of coverage and longer length associated with the liner tube it is ideal for garden centre applications which often comprise of long greenhouse type structures.

Figure 7.6 Linear tube radiant heater

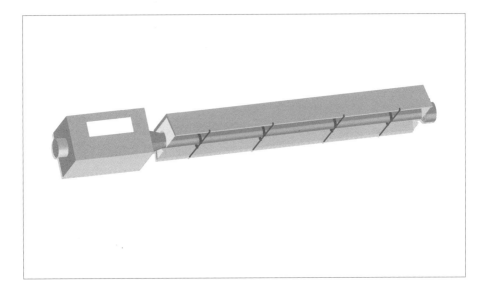

Figure 7.7 Linear tube (multi-burner)

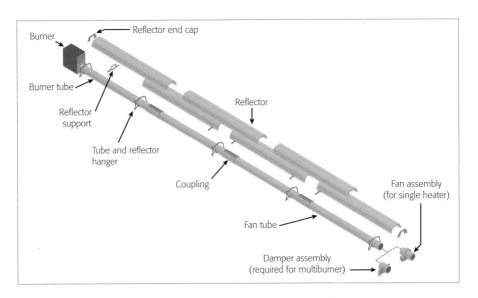

Figure 7.8 Continuous tube heater

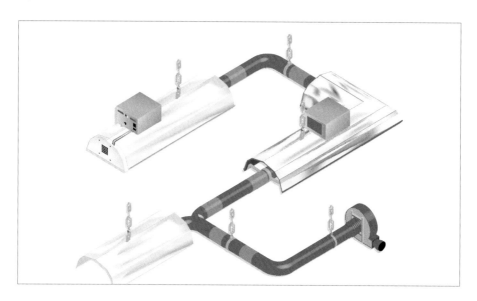

Linear tube (multiple burner)

These types of linear systems comprise of an assembly of multiple linear radiant tube units.

However, as can be seen from Figure 7.7, instead of utilising individual fans for each tube, they collect to a common fan unit, which will be suitably sized by the manufacturer to exhaust flue gases from the total number of burner units in the assembly.

These systems will be purpose designed by the manufacturers to meet bespoke heating system requirements for a given application and can therefore be built-up to provide individual zone heating within the overall scheme. Each zone could have its own individual programmable time-switch and thermostat control.

Continuous tube

Continuous systems are generally known as 'CoRayVac' or 'NorRayVac' systems, these are manufacturer's trade names for the system.

Floor area coverage is provided by long runs of the heat treated aluminised steel tubing suspended from the roof structure of the building. Reflectors are provided above the tubes to direct heat downwards (see Figure 7.8).

The system utilises an incremental burner system whereby several burners are located in series along a single tube heat exchanger, which carries the combustion gases from preceding burners in the system.

Burners are located approximately every 3.5m to 6m along the system, dependent on the rating of the individual burners and the required radiated intensity level. Generally, lower burner rates will be used for lower mounting heights.

Figure 7.9 Schematic of a typical continuous heater installation

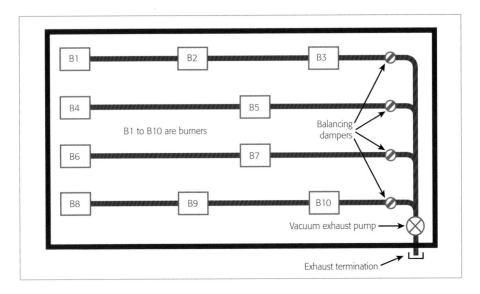

The firing rate of each burner can be individually set and the distance between burners can be determined at the design stage, so that different radiated intensity levels are possible at various sections along the system, to match heat losses from each area within the building.

For example, consider a system in a work space which has 4 radiant tube branches feeding into a common header (illustrated in Figure 7.9). The outer tubes could be set at a higher radiated intensity to combat the heat losses from the walls compared to the inner legs which would actually have heat gain from the outer tubes.

Similar to the linear system with multiple burners, previously described, the continuous system incorporates a single vacuum pump to draw all POC along the tube and exhaust them to atmosphere at the end of the tube system.

A balancing damper will be installed at the end of each leg to correctly set and balance operating vacuums along each tube.

Air heated large bore tube

This system has been used in the past for larger heat inputs and incorporates a floor standing direct-fired air heater with its ducted hot air output connected to a large bore tube, or system of multiple tubes, installed above the area to be heated and suspended from the roof support steelwork of the building.

The tubes are insulated on their top side to minimise heat loss above and are shielded at the side to prevent heat loss due to cross-flow of air currents. Tube diameter ranges nominally from 230mm to 600mm.

The heater provides hot air at approximately 200°C through the tube which is then exhausted to atmosphere at the end of the tube system, much like an oversized 'U' tube heater. The average surface temperature emitted from the tube will be approximately 150°C.

Heat output will be measured per metre run of tube and can range from 0.5 – 5kW/metre, maximum heat input nominally 600kW.

The relative merit of this system was that it provided a more even spread of heat, due to the lower surface emitted temperature, over a larger floor area. The main disadvantage of the system was that of initial high capital cost.

Typical applications for the system would be warehouses, bus stations, aircraft hangars, large garage areas and so on.

Because the installations were of a specialist nature, they were only marketed by a few companies, which undertook the design, manufacture and installation of the whole assembly.

Catalytic panels

Many people will be familiar with the catalytic panel heaters for use with LPG on touring caravans etc. Similar heater panels are available for use on NG, but are usually found on process equipment used in the infrared drying of manufactured products.

Processes such as paint drying or leather drying involve the use of solvents or gases with a relatively low ignition point. A catalytic panel will dry the product and reduce the risk of explosion associated with the volatile nature of solvents etc.

The principle of operation utilises fuel gas supplied to a porous refractory panel, which is impregnated with a platinum based catalyst. The fuel gas will react to the catalyst, which will oxidise the gas and produce heat at approximately 450°C without the presence of a flame.

Installation requirements

The following information provided in this Part offers general guidance for the installation of the types of heaters previously described.

Specific information relating to the installation of any particular heater or system will be described in the appropriate manufacturer's instructions.

Particular installation normative reference documents include:

- BS 6896: 2011 Specification for installation and maintenance of gas-fired overhead radiant heaters for industrial and commercial heating (2nd and 3rd family gases)

- BS EN 13410: 2001 Gas-fired overhead radiant heaters. Ventilation requirements for non-domestic premises

- IGEM/UP/2 (Edition 2) Installation pipework on industrial and commercial premises

- IGE/UP/10 (Edition 3) with amendments October 2010 Installation of flued gas appliances in industrial and commercial premises incorporating specific requirements for appliances fired by bio-fuels

- BS 5440-1: 2008 Flueing and ventilation for gas appliances of rated input not exceeding 70kW net (1st, 2nd and 3rd family gases) – Part 1: Specification for installation for gas appliances to chimneys and for maintenance of chimneys

- EH40/2005 Workplace Exposure Limits Containing the list of workplace exposure limits for use with the Control of Substances Hazardous to Health Regulations (as amended) – available freely as a pdf download from www.hse.gov.uk/pubns/books/eh40.htm or as a hard copy from HSE Books

When designing any radiant heating system there are many areas of consideration; collaboration will be essential between those involved such as the gas operative, the occupier and the building owner.

There may also be the need to inform local authorities for appropriate planning consent (e.g. for listed buildings), fire authorities, building insurers etc. Additionally, the installation will need to take in to account the requirements of Building Regulations applicable to the geographical area concerned (see **Part 2 – Gas and associated legislation** for further guidance).

Matters to be considered include:

- heater manufacturer's installation requirements

- building construction (fabric heat losses, suitability and proximity of materials etc.)

- flueing requirements

- ventilation requirements

- gas availability
(type, site capacity and pressure)

- electricity requirements

- site access and liaison with associated trades, etc.

Note: The above list should not be regarded as exhaustive.

Competence

Work carried out relative to the installation of overhead radiant heating systems needs to be undertaken by operatives who are experienced in such work and who hold the relevant certificates of competence obtained through the ACS and/or aligned N/SVQ's.

Note: For more information regarding commercial ACS categories refer to Part 3 – 'Competence' and the current Essential Gas Safety – Non-Domestic.

Appliance suitability

Any appliance or system must be fit for the purpose for which it is intended. This can be ascertained from a variety of sources:

- CE mark – ensure the appliance carries a CE mark together with the identification number of the notified body responsible for Quality Control (QC) surveillance

- manufacturer's data – check that appropriate instructions are available and that the appliance displays a data plate with essential information – maximum input gas pressure; burner gas pressure; kW heat input rating, etc.

  - the data plate should also display:

    - country of import (e.g. GB)

    - type of gas and supply pressure (e.g. G20 – Natural gas, G30 – Butane and G31 – Propane)

    - heater/multi burner system category as defined in BS EN 437: 2003 + A1: 2009 'Test gases. Test pressures. Appliance categories'

General location requirements

Any radiant heating system will need to be installed, such that persons occupying the heated zone are not subjected to excessive levels of radiation – maximum radiation level at head height (2m) should not exceed 240W/m².

If this level is exceeded, then it may cause occupants to feel unwell, in much the same way as excessive exposure to sunlight can.

Radiant heaters will need to be sited clear of combustible materials e.g. fabric of the building, particularly with some older structured buildings where roof support beams may be of timber construction.

Manufacturers will normally provide guidance on minimum distances from combustible material.

Appliance manufacturer's information will also provide clearance requirements for adequate maintenance access.

All items that are liable to wear or otherwise require regular attention such as burners, fans and control devices will need adequate access provision to allow withdrawal/removal without causing damage to other components on the unit or system.

In factory and warehouse buildings there is likely to be significant movement of fork lift trucks, or other hoisting mechanisms. Any heater units will need to be located with due consideration of such equipment to avoid any possibility of heater damage. Furthermore, where there is a perceived risk of physical damage to the heater units, it may be necessary to provide some form of mechanical barrier in order to minimise risk.

When radiant tube heaters are located in sports halls there is a risk of the unit being regularly hit by balls (footballs, tennis balls, etc). Not only could this damage the heater, but also put a strain on the heater support mechanism. Wall mounted support brackets for instance may need to be secured more substantially as a result.

Smaller projectiles such as badminton shuttlecocks can become lodged adjacent to the appliance causing them to burn or ignite; manufacturers may provide accessories such as purpose made guards to combat this problem.

Restricted locations

Individual heaters and multi burner systems will need to be installed in locations that are deemed suitable. In industrial premises particularly, there will be a wide variety of applications and processes, therefore, due consideration needs to be given to the likely environmental conditions:

- corrosive or salt-laden atmospheres will obviously have an affect on metallic components, particularly burners and controls, e.g. metal plating/treatment workshops

- dusts and vapours given off from plastic forming processes, cleaning or curing applications when passed into the atmosphere will be drawn into the combustion chamber and may subsequently produce harmful gases

- chlorine laden atmospheres (e.g. at swimming pools/baths) – when atmosphere is laden with chlorine in the vicinity of a permanent heat source, such as permanent gas pilot light, the chlorine will break down to base component chemicals, including acids which will attack the metallic components of the heater and corrode very rapidly.

- high velocity air movement in the vicinity of the burner will affect its performance and may prevent complete combustion from taking place

- petroleum or heavier than air flammable vapours represents a significant risk of explosion. Therefore, heaters or multi-burner systems must not be installed in areas that are classified as being hazardous in accordance with BS EN 60079-10-1: 2009 'Explosive atmospheres. Classification of areas. Explosive gas atmospheres'.

Where the area is not classified as being hazardous but heavier than air vapours might be present – as a result of spillage at a vehicle workshop/garage for example – any gas heater located in such an area, will need to be sited so as not to represent an ignition source to the fuel spillage i.e. the base of the heater will need to be mounted at least 1.8m above floor level.

In addition, electricity supply cables and switchgear will need to be sited above 1.2m or appropriately protected as defined in BS EN 60079-10-1.

Note: For additional guidance for garages, refer to IGEM/UP/18 'Gas installations for vehicle repair and body shops'

BS EN 60079-10-1 defines areas in distinct zones, i.e. zone 0, 1 and 2, relative to known hazards that may be present. Increasing the amount of ventilation in a hazardous area may in certain cases change the zone rating of an area and may then permit installation of gas-fired heater units.

- LPG fuelled appliances cannot be located in a totally enclosed room below ground e.g. cellar or basement.

If there is any doubt as to the suitability of any heating system for a given application, always seek advice from the appliance manufacturer(s).

Most manufacturers will prefer to give, often free advice in order to ensure the heating is appropriate for the application, rather than risk the possibility of inheriting bad publicity resulting from unsuitable/poor installations.

Appliance support/mounting

Individual radiant plaques or tubes may be bracketed to a suitable wall structure. Wall mounting brackets are usually available as an optional extra from the manufacturer.

The brackets will need to support the heater at an angle of inclination between 30° and 45°. The angle can usually be varied by adjusting the drop rods or chain on each bracket (see Figure 7.10).

Note: When installing a tube heater at an angle, the combustion fan outlet should be positioned so that it remains vertical.

The location of a particular heater or multiple heater system may be dictated by the need for adequate support or suspension facility. The structure of the building may restrict possible locations for this facility.

Any support or suspension facility will need to be of sufficient number and strength to support the full weight of the heater system. Also the particular means of support will need to be constructed of suitable materials not liable to corrode.

The building structure will need to be sufficiently strong to withstand the extra load placed upon it by the heaters or heater system complete with the associated flue system, particularly with multiple burner linear or continuous systems.

In the event of suitable roof steelwork not being available, additional steelwork should be fitted to enable hangers to be used for suspending the heaters.

If there are any doubts as to the strength or suitability of building structure to which heaters or system is to be suspended, reference will need to be made to a building consultant/architect/structural engineer.

Attachment to the heater support lugs should be made by either, a 'D' shackle, nut/bolt and large washers, or in the case of drop rods, a closed formed hook (see Figure 7.11).

The hanging attachments to overhead steelwork etc. need to be purpose made to good sound engineering practice or of a proprietary type fixing.

Note: Continuous or multiple burner linear tube systems will require a downward slope, away from the burner to ensure that any condensation will cascade away from the burner towards the exhaust fan, where a suitable condense trap will be required (see Figure 7.12).

Air supply/ventilation

To achieve complete combustion, it is essential to ensure that the appliance is supplied with an adequate supply of clean fresh air (ventilation), in accordance with the manufacturer's instructions.

This will necessarily take into account the type of flue fitted to the appliance and any other appliances in the same area, along with any cooling requirements for the appliance and the occupants of the workspace in which the appliance is sited.

Figure 7.10 Typical angle bracket detail

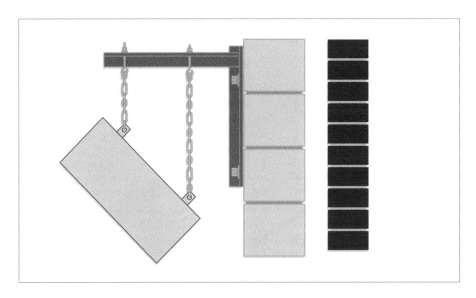

Figure 7.11 Examples of heater support lugs

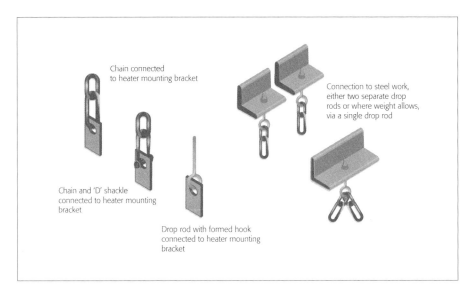

Figure 7.12 Example of tube suspension assembly

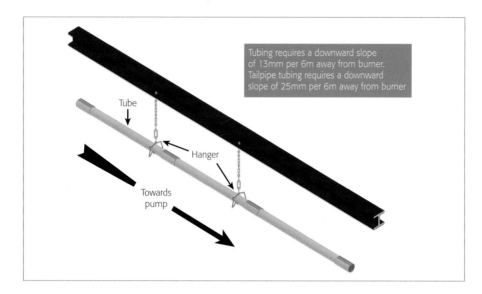

Note: For information relating to the general ventilation requirements in non-domestic establishments, see Essential Gas Safety – Non-Domestic (Order ref: ND1, see Part 13 – CORGI*direct* Publications).

In addition to manufacturer's instructions, the industry standards referred to for ventilation installations specific to radiant heating systems are as follows:

- BS 6896

- BS EN 13410

- IGEM/UP/10 (Edition 3)

- BS 5440-2: 2009 Flueing and ventilation for gas appliances of rated input not exceeding 70kW net (1st, 2nd and 3rd family gases) – Part 2: Specification for installation and maintenance of ventilation provision for gas appliances.

Gas installation

The GSIURs require gas fitting operatives to ensure that gas installation pipework and fittings are installed safely with due regard to the siting of other services e.g. other pipes; pipe supports; drains; sewers; cables; conduits and electrical control equipment.

Gas operatives will also need to be aware of the limitations of the building structure when installing gas equipment and pipework.

Note: For detailed information relating to the general requirements for gas pipework in non-domestic establishments, see Essential Gas Safety – Non-Domestic.

The industry standard referred to, generally for non-domestic pipework installations is IGEM/UP/2 (Edition 2).

Gas connection

Manufacturers usually stipulate that the final connection to the burner/heater unit is by means of a suitable gas flexible pipe.

The flexible pipe material being stainless steel to BS 6501-1: 2004 'Metal hose assemblies. Guidance on the construction and use of corrugated hose assemblies' and minimum acceptable quality 'type B class 1'.

For added protection it is advisable though not mandatory that the flexible pipe connector incorporates a surface cover. This may be a plastic sleeve, or if there is risk of physical damage, over-braided stainless steel.

Alternatively, it is acceptable to use a gas 'Caterflex' which is a plastic coated stainless steel flexible pipe manufactured to BS 669-2: 1997 'Flexible hoses, end fittings and sockets for gas burning appliances. Specification for corrugated metallic flexible hoses, covers, end fittings and sockets for catering appliances burning 1st, 2nd and 3rd family gases' and incorporates a self-sealing plug and socket.

Any gas flexible pipe needs to be installed such that there are no sharp 90° bends and the flex is not subjected to stress or torsion.

A slow radius bend of 180° onto the burner will usually be required, to take up expansion and contraction created at the heater by its heating and cooling cycles.

Where a self-sealing plug and socket is not used, a union joint will be required at one end of the flexible pipe to facilitate disconnection of the gas pipe for maintenance of the burner/heater unit. Installation should ensure that the flexible does not twist when the union is tightened.

In any event, a 90° operable gas isolation valve will need to be installed immediately upstream of the flexible pipe assembly.

Due consideration should be given to the available gas supply pressure in the pipe system feeding the heater.

On some industrial and commercial sites it may be possible for the metered supply to have elevated gas pressure, or on large sites, such wide variation on load characteristics as to create significant variation in the gas line pressure.

Where either of these situations exist, it will be necessary to check with the appliance manufacturers to establish if there is a need for fitting an additional gas regulator upstream of the appliance control system.

In certain cases, particularly where the supply pressure to the appliance is 75mbar or more, there will need to be adequate automatic means for preventing the appliance and associated pipework from being subjected to this higher pressure e.g. in the event of regulator diaphragm failure.

This would usually be by means of a regulator with integral overpressure slam-shut valve incorporated.

Electrical connections

Non-domestic properties normally utilise three-phase electrical systems due to the high current demands of the electrical equipment installed and used.

When gas operatives are required to work on gas appliance/equipment, which utilise either three-phase or single-phase electricity, it is important that those gas operatives are competent and have the relevant knowledge and experience to work safely on those systems and be able to carry out a thorough risk assessment of any potential hazards that may be involved.

For further information on risk assessments see Essential Gas Safety – Non-Domestic – Part 3 – 'Risk assessments' for further guidance.

Important: The Electricity at Work Regulations 1989 (EWR) apply to any operative carrying out any type of electrical work. It is a requirement of the EWR that the person is competent to undertake the work.

Where any doubt exists with regards to competency or lack of knowledge/experience to work on those electrical systems, or where a risk assessment identifies that the hazard would be unacceptable, no work should be undertaken.

Working on three-phase equipment, supplies and four wire supplies

In the UK the 400V, 4-wire three-phase and neutral (TPN) method is widely used for the distribution of supplies within commercial and small industrial installations.

For example, on large wet central heating systems the circulating pump may operate at 400V from a three-phase supply while other components such as thermostats and time switches may be 230V single-phase or sometimes even less, for example 24V, which is classed as Extra Low Voltage (ELV).

On commercial warm air heaters, the fans (combustion and/or circulating) may be three-phase, while other components may be single-phase.

It must be remembered that each of these phases is carrying 230V. A balanced three-phase circuit does not require a neutral for it to be able to operate. This is because there is always a return path for the current through one of the other two phases.

There are a number of signs to look for when trying to identify the voltage on non-domestic gas installations, two most basic signs are:

1. Appliance data plate – one of the simplest methods of determining the voltage to a non-domestic gas appliance is by checking the information on the appliance data plate, assuming of course a plate is present.

   The information displayed will typically consist of:

   - supply voltage – 400V
   - supply frequency – 50Hz
   - no of phases – 3
   - no of wires – 4
   - power consumption – 4.0kW
   - fuse rating – 400V, 25A HBC
   - fuse rating – 230V, 5A

2. Installation wiring – all installation wiring falls within the scope of BS 7671: 2008 (Incorporating amendment No. 1) 2011: 'Requirements for electrical installations. IET Wiring Regulations Seventeenth Edition' and therefore requires completion of electrical certification when installed.

   The existence of a three-phase supply can often be confirmed by the wiring colour code, this will need to be tested to ensure it is correctly sized etc. before undertaking any electrical installation work.

Remember, all electrical installations should be in accordance with BS 7671 and the supplied manufacturer's wiring guides, especially for the correct connection of supplies.

Of equal importance is the correct routing of cables, including suitable supports, where required and the correct securing of cables to individual pieces of equipment.

Each appliance and its associated controls needs to have its own means of electrical isolation, which should be situated in close proximity to the equipment it serves and have suitable means for the application of securing devices, i.e. hasps or padlocks for ensuring that the appliance cannot be re-energised whilst being worked on by an operative (this should be highlighted as part of a risk assessment).

Chimney systems

Appliances and their flues are classified by a 'European Committee for Standardization' document entitled 'PD CEN/TR 1749: 2009 'European scheme for the classification of gas appliances according to the method of evacuation of the products of combustion (types)'.

The purpose of this document is to harmonise, across Europe, the classification of appliances burning combustible gases.

PD CEN/TR 1749 separates gas appliances into three types:

* Type A – flueless

* Type B – open-flued (including systems formerly known as 'closed' flue)

* Type C – room-sealed.

For further detail on the correct flue classification for general appliance types, see Essential Gas Safety – Non-Domestic (Order Ref: ND1, see **Part 13 – CORGI***direct* **Publications**).

The classifications applicable to radiant tube heaters are as follows:

* Figure 7.13 – Type B_{12} heater, is a type B_1 heater designed for a natural draught flue, incorporating a fan downstream of the heat emitting tube and upstream of the down draught diverter.

* Figure 7.14 – Type B_{13} heater, is a type B_1 heater designed for natural draught flue, incorporating a fan upstream of the heat emitting tube and draught diverter.

* Figure 7.15 – Type B_{22} heater, is a type B_2 heater (no draught diverter), incorporating a fan downstream of the heat emitting tube.

* Figure 7.16 – Type B_{23} heater, is a type B_2 heater (no draught diverter), incorporating a fan upstream of the heat emitting tube.

* Figure 7.17 – Type C_{12} heater, is a type C_1 heater (horizontal balanced flue), incorporating a fan downstream of the heat emitting tube.

* Figure 7.18 – Type C_{13} heater, is a type C_1 heater (horizontal balanced flue) incorporating a fan upstream of the heat emitting tube.

* Figure 7.19 – Type C_{32} heater, is a type C_3 heater (vertical balanced flue), incorporating a fan downstream of the heat emitting tube.

* Figure 7.20 – Type C_{33} heater, is a type C_3 heater (vertical balanced flue), incorporating a fan upstream of the heat emitting tube.

Change to terminology

European standardization has also brought about changes to terminology typically used within the gas industry. One such change is with regards to what we in the UK have referred to as 'flues'.

Under European terminology 'flue' describes the passage of the POC and not the material or structure for the transportation of that POC. Where we in the UK have used the term flue to also describe the material/structure, this is defined as either 'chimney' or 'chimney component'.

Therefore and in the main, this manual will use the newer term (chimney) when describing flue construction/material as well the term 'flue pipe' (connecting and appliance to a chimney), as appropriate and 'flue' (when describing the passage of POC).

Note: Current Standards use a mixture of terms, including the older term of flue and therefore, where used will still be relevant to that Standard.

Figure 7.13 Type B$_{12}$ heater

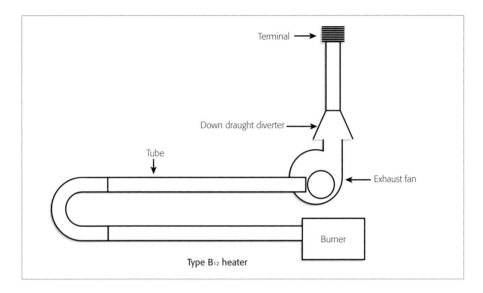

Type B$_{12}$ heater

Figure 7.14 Type B$_{13}$ heater

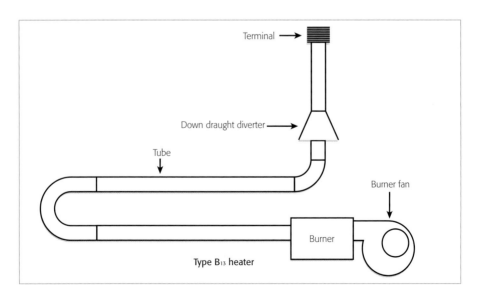

Type B$_{13}$ heater

Figure 7.15 Type B$_{22}$ heater

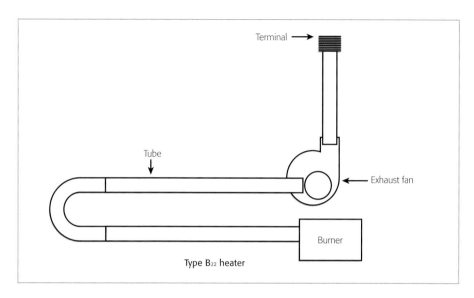

Terminal

Tube

Exhaust fan

Burner

Type B$_{22}$ heater

Figure 7.16 Type B$_{23}$ heater

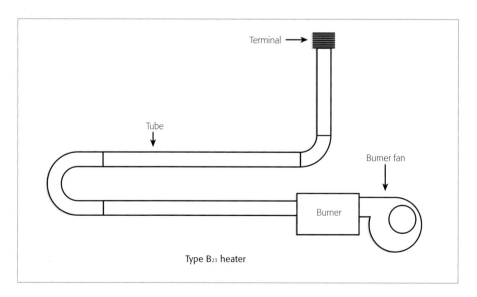

Terminal

Tube

Burner fan

Burner

Type B$_{23}$ heater

Figure 7.17 Type C$_{12}$ heater

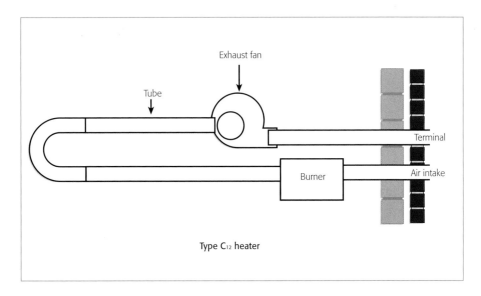

Type C$_{12}$ heater

Figure 7.18 Type C$_{13}$ heater

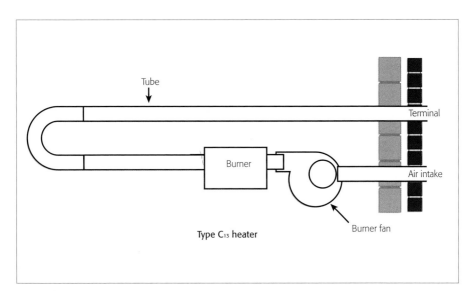

Type C$_{13}$ heater

Figure 7.19 Type C$_{32}$ heater

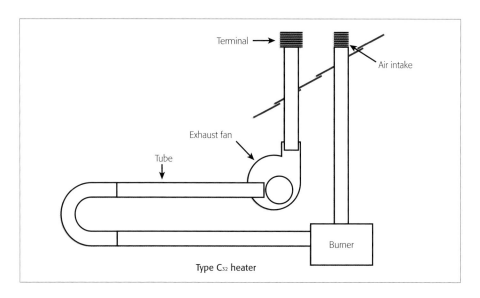

Type C$_{32}$ heater

Figure 7.20 Type C$_{33}$ heater

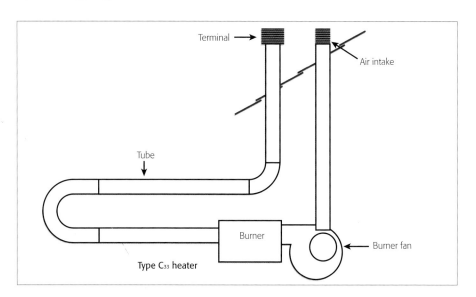

Type C$_{33}$ heater

Figure 7.21 Typical flue adaptor detail

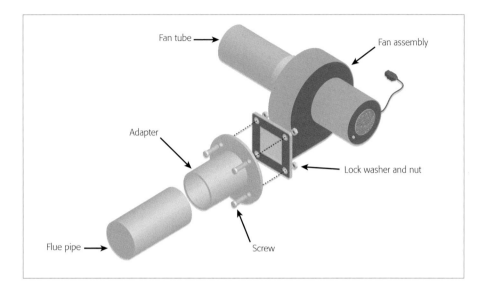

Chimney installation

Any chimney/flue pipe system will need to be installed in accordance with manufacturer's instructions to ensure safe and complete evacuation of combustion products to the atmosphere.

Chimney components and their materials of construction will need to be in accordance with the appropriate British Standards.

Note: For details on the general appliance chimney/flue pipe installation, see Essential Gas Safety – Non-Domestic (Order Ref: ND1, see Part 13 – CORGI*direct* Publications).

Single heater installation

Unless otherwise stated in manufacturer's instructions, it will be necessary to install the chimney/flue pipe so that it is supported independently of the heater unit.

The manufacturer's chimney adaptor piece should always be used to ensure correct connection to the tube/fan housing and a complete seal is obtained (see Figure 7.21).

The adaptor piece may also allow disconnection where necessary, for proper access to the fan, to facilitate adequate maintenance.

The fan outlet can usually discharge in either the vertical or horizontal plane, but if the chimney/flue pipe is likely to be in excess of 5m long, then dependent upon heat input rating, it may need to be insulated.

Figure 7.22 'Herringbone' system

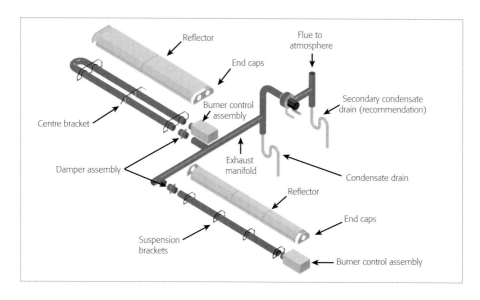

Check with the manufacturer, some allow longer lengths of single wall flue pipe for larger heat inputs, e.g. 25kW input heaters could allow up to 8m in some cases. However, if there any 45° bends in the chimney/flue pipe system this will affect its performance and again manufacturers should be consulted.

For Type B and C systems, the flue terminal will need to be positioned so as to ensure that no combustion products are allowed to re-enter the building or re-circulate into the combustion air intake. If the terminal is located less than 2m above ground level, then it needs to be fitted with a suitable guard to prevent possible blockage or injury to persons.

Any room-sealed flue system will need to be assembled and installed with manufacturer's approved components.

Current standards require that a data plate is displayed on existing chimney/flue pipe systems detailing the various performance characteristics.

Where a heater is to be connected to such an existing chimney/flue pipe, the installer will need to verify with the heater manufacturers that the information displayed on the chimney/flue pipe data plate is appropriate for the connection of the proposed heater(s).

Multiple heater installation

Heaters and multi-burner systems can only be installed on a common chimney if the system is approved by the heater manufacturers and complies with their installation instructions.

Such a system is known as the 'herringbone' system illustrated in Figure 7.22. The illustration shows the application is appropriate for 'U' or 'linear' tube heaters.

Figure 7.23a Flue damper arrangement

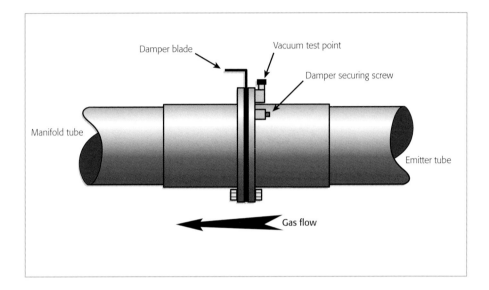

Figure 7.23b Flue damper arrangement

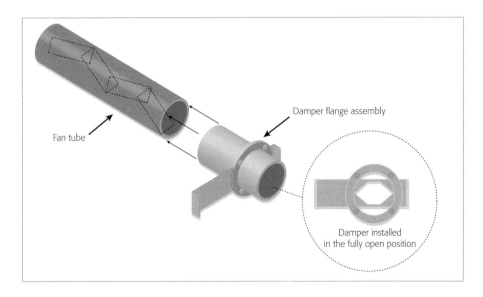

Figure 7.24a Horizontal flue outlet

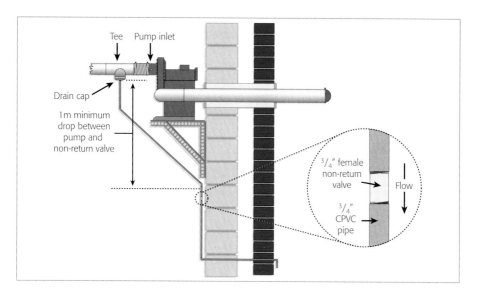

Due consideration will need to be given to the possibility of condensation formation. If this is likely for periods in excess of the first hour from start-up, provision should be made to ensure condensate is removed. Preferably by means of a condensate drain of at least 22mm diameter.

In addition, chimney/flue pipe material will need to be insulated, e.g. a double walled flue system, with inner lining resistant to corrosion caused by condensates.

In order to balance the system correctly, suitable dampers where the heaters connect into the exhaust manifold, will usually need to be installed. Where dampers are fitted, they need to be installed with appropriate interlocks or other safety feature, so as to comply with manufacturers instructions, standards and, where appropriate, regulations.

For various manufacturer's details of damper arrangements refer to Figure 7.23 (a) and (b).

Continuous tube systems

The chimney/flue pipe installation on a continuous tube system is connected to the outlet of the exhaust fan system. Figure 7.24 (a) and (b), show typical arrangements for horizontal and vertical systems.

As will be noted from the systems detailed, an important part of the chimney/flue pipe system is the facility to discharge condensation.

Condensation will be formed as a result of the long lengths of heat emitting tube, along which the combustion products must pass to the exit point.

This represents a long length of cold steel tube which must be heated; inevitably there will be times whereby the POC will be below the due point of the gases (55°C – 60°C) at sections of the system, particularly on zone controlled tube sections.

Figure 7.24b Vertical flue outlet

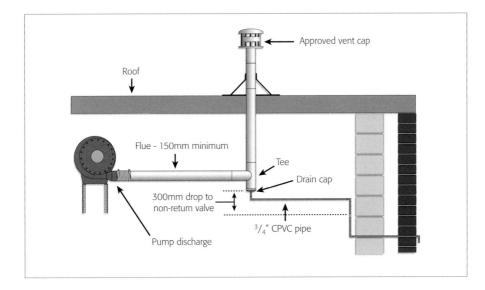

The heat emitting tube will also need to be installed, such that there is a continuous incline in accordance with manufacturer's instructions. This is to ensure that any condensation formed, drains towards the drain cap.

With continuous tube systems, the cumulative kW rating of the system can be in excess of 150kW (gross). In which case, reference should be made to the third edition of the Clean Air Act 1956 Memorandum – to determine specific requirements for chimney heights.

Also, in some cases, the height and termination of the chimney may need Local Authority approval. A report would then be required providing details on chimney commissioning including the emissions data to be submitted to the Local Authority Environmental Health Officer.

A copy of the relevant information will need to be left on site also.

Commissioning

Manufacturer's instructions will provide specific information for commissioning any particular appliance and its associated equipment.

Additionally, IGE/UP/4 (Edition 2) 'Commissioning of gas-fired plant on industrial and commercial premises', offers further generic guidance for the commissioning process.

Correct commissioning of an appliance in accordance with the manufacturer's instructions, is as critical to a safe and efficient installation, as its initial installation.

Therefore, where the installation can not be commissioned immediately after installation, it should be isolated from the gas supply and suitable notification attached (this fact also needs to be documented on any appropriate paperwork) to advise the reader that the installation is un-commissioned.

The following is provided for information purposes only and should not be regarded as substitute for the source documents.

Planning

Before travelling to site, a certain amount of preparation will be necessary, including:

- ensure all relevant manufacturer's information is available

- all relevant drawings and system plans are available

- all gas tightness test/purge certificates have been completed by the upstream pipe installer

- all electrical test certificates have been completed by the electrical installer

- risk assessment has been completed

- COSHH statements prepared, where relevant

- any hot work or permit to work has been granted

- access facilities are available. Due to the fixing height of this type of heater, suitable scaffolding or hydraulic lift facility (cherry picker) will usually be required

- all necessary tools are available, are suitable for the purpose and are of sound quality. Any electronic instruments need to be correctly calibrated and certificated accordingly

- a written plan of required work is completed (usually provided in manufacturer's instructions).

Inspection

Once on site, a complete inspection of the installation will need to be carried out to ensure that:

- all gas and electrical supplies are suitably sized, correctly located and supported and that the installation has been installed in accordance with the manufacturer's instructions and relevant standards

- assembly is complete and all components are fit for the intended purpose

- the appliance has been correctly supported/suspended, it is level/stable with adequate clearances for maintenance and from combustible materials

- gas and electrical supplies have been isolated.

Activation – dry run

Subject to satisfactory inspection, the next phase of the commissioning process is to carry out preliminary physical checks with the fuel isolated:

- a suitable gas tightness test is undertaken on the appliance connector pipe

Note: For further detail on the correct procedures for testing gas pipework in non-domestic establishments, see Essential Gas Safety – Non-Domestic (Order Ref: ND1, see Part 13 – CORGI*direct* Publications).

The industry standard referred to generally for non-domestic pipework testing and purging is IGE/UP/1 (Edition 2) 'Strength testing, tightness testing and direct purging of industrial and commercial gas installations' – or alternatively if within its scope, IGE/UP/1A (Edition 2) 'Strength testing, tightness testing and direct purging of small, low pressure industrial and commercial natural gas installations'.

- check that all manual isolating valves and safety shut off valves are closed and are gas tight (e.g. no let-by)

Figure 7.25 Checking cold suction

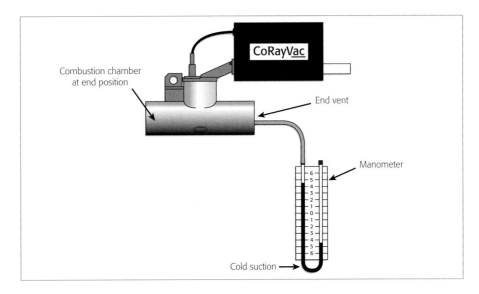

- any control interlocking device is set to its provisional operating level, considered safe for commissioning (e.g. regulator, process controls and interlocks)

- in the case of common chimney systems, or continuous tube systems, the balancing dampers will need to be in the open position initially. The required level of 'cold' vacuum can then be checked and dampers adjusted starting from the tube end furthest from the exhaust (see Figure 7.25).

- with auxiliary energy supplies available, all power equipment and interlocks are checked for operation, e.g. where automatic control units are employed, the controller is checked at each stage of its light-up sequence for correct operation and response of the control unit

- check that the system of ignition is adequate, e.g. strength and location of any spark generation is correct to ensure smooth and reliable lighting of start gas flame

- flame safeguard systems correctly go to shutdown condition in the absence of a flame, e.g. lockout within the specified time period. If a simulated flame is present the system detects the presence of the flame correctly.

Activation – fuel run

Once all checks have been made with the fuel isolated to the satisfaction of the commissioning operative, the gas may be turned on and the following further checks initiated:

- the appliance is correctly purged of air (pipework connector and combustion chamber)

- allow the burner system to proceed through to start gas ignition or pilot stage, ensure that start gas flame or pilot is established and set at the correct pressure before proceeding to main gas

- confirm by suitable means of leak detection that all start gas or pilot gas pipework is gas tight

- ensure that loss of start gas/pilot flame, results in the correct shutdown of the control system. Check thermo-electric devices for the correct 'drop-out' time as described in manufacturer's instructions

- ensure that, with the burner shutdown, the safety shut off valve(s) remain gas tight

- with main gas available and controls restricted to a nominal main flame ignition rate, allow the burner to proceed to main gas ignition. Ensure that main gas flame is established and set at the correct pressure. Continuous tube systems use zero pressure regulator technology and so the regulator itself is preset by the manufacturers, it will not usually require adjusting. The correct gas rate is established by varying the vacuum in the tube (see **Operational checks**)

- confirm by suitable means of leak detection that all main gas burner pipework is gas tight

- ensure that loss of gas flame results in the correct shutdown of the control system

- ensure that, with the burner shutdown, the safety shut off valve(s) remain gas tight.

Operational checks

Once all checks have been made to ensure safe and reliable ignition to main gas burner stage, (to the satisfaction of the commissioning operative) the following further operational checks can be initiated:

- the heating plant is allowed to run up to normal operational temperatures to ensure the unit remains satisfactory

- operational controls can now be checked for satisfactory operation, e.g. room thermostats

- for common chimney systems and continuous tube systems, the 'hot' suction can now be checked in a similar manner to checking the cold suction. It may sometimes be necessary to make adjustments to balancing dampers or even replace end plates to achieve the required vacuum. See Figure 7.26 for typical end plate/vent detail, different plates are available from manufacturers, with different vent arrangements to increase/decrease vacuum in the tube branch

- combustion gases are effectively dispersed, ventilation is adequate

- combustion checks are carried out. For flueless appliance systems (Type A), it will be necessary to ensure that POC do not build-up beyond accepted limits within the heated environment

- undertake spillage tests, where appropriate

- carry out essential checks required by the GSIUR i.e. the effectiveness of the flue, the supply of air, gas pressure and gas rate and the general safety of the appliance.

BS 6896, Annex A 'Testing for carbon dioxide in the environment', describes the maximum CO and CO_2 concentrations at any point where persons can normally be expected to work. This includes low levels, mezzanine levels and at high levels.

At these points CO_2 concentration should not exceed 2800 ppm and CO limited to 10ppm.

Checks should be undertaken when the building is operating under full heat input conditions, usually after at least the first hour of operation.

In the case of spot heating applications it will normally only be necessary to check the environment in the immediate vicinity of the heater unit.

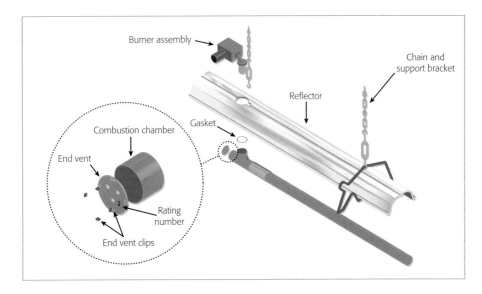

Figure 7.26 Typical end plate/vent detail

Burner assembly

Chain and support bracket

Reflector

Combustion chamber

Gasket

End vent

Rating number

End vent clips

The HSE provide further guidance on exposure limits for various substances that can be encountered in the workplace via EH40/2005 'Workplace exposure limits'.

Workplace Exposure Limits (WELs) are provide for short term exposure (15 mins) and for long term exposure (8hr time weighted average (TWA), i.e. to cover a typical working shift), which should not be exceeded.

The WELs for POC are:

- CO – 30ppm (8hr TWA) & 200ppm (15 mins)

- CO_2 – 5000ppm (8hr TWA) & 15000ppm (15 mins)

To undertake these tests, the correct instrumentation will need to be used to ensure it is suitable for the application and correct levels of accuracy are achieved.

For example, it will not usually be appropriate to use typical portable electronic combustion gas analysers (ECGAs) to sample CO_2 in the atmosphere - those analysers complying with BS EN 50379-4: 'Specification for portable electrical apparatus designed to measure combustion flue gas parameters of heating appliances. Performance requirements for apparatus used in non-statutory servicing of gas-fired heating appliances' – as they do not measure CO_2 directly, but merely calculate a CO_2 level from the directly related excess oxygen measured in the sample.

Where measuring CO_2 directly, the ECGA should comply with BS 8494: 2007 'Electronic portable and transportable apparatus designed to detect and measure carbon dioxide in indoor ambient air. Requirements and test methods'.

Note: For further detail on the correct procedures for combustion analysis and appliance efficiency testing, see 'Essential Gas Safety – Non-Domestic' (Order Ref: ND1) and 'Combustion performance testing - non-domestic' pocket guide (Order Ref: CPA2), see Part 13 – CORGI*direct* Publications.

The industry standard referred to generally for using ECGAs in non-domestic premises is BS 7967-5: 2010 'Carbon monoxide in dwellings and other premises and the combustion performance of gas-fired appliances. Guide for using electronic portable combustion gas analysers in non-domestic premises for the measurement of carbon monoxide and carbon dioxide levels and the determination of combustion performance'.

Completion

Once all aforementioned checks have been made to the complete satisfaction of the commissioning engineer the following will be required:

- all users are instructed in the correct operation of the heater and/or system and its user controls, light-up and shutdown sequence

- manufacturer's 'user' instructions are left for the responsible person. The 'responsible person' is defined by the GSIUR as the 'owner or occupier'. However, on many large commercial or industrial sites this may not be relevant, it may be more appropriate to identify the works engineer or manager for this purpose

- the responsible person should also be advised concerning matters such as actions to be taken in the event of fault or emergency conditions and advice concerning regular maintenance of the units

- a suitable report will need to be completed and left with the responsible person. This report needs to detail the final setting parameters of the heater/system, including:

  - customer/site details

  - plant/heater details, e.g. make, model and serial number

  - fuel supply details, e.g. type and supply pressure

  - appliance/system operating set levels

  - combustion and emission data

  - electrical data, e.g. nominal supply, overload settings, fuse ratings.

Note: CORGI*direct* produces suitable forms for the correct documentation of checks/tests carried out, which gas operatives can provide to their customers. The forms, 'Plant Commissioning/Servicing Record' (Order Ref: CP15), 'Gas Testing and Purging' (Order Ref: CP16) and 'Gas Installation Safety Report' (Order Ref: CP17) which carry the Gas Safe Register logo under licence are detailed in Part 13 – CORGI*direct* Publications.

Handover

Upon satisfactory completion of all the aforementioned commissioning actions, the commissioning report, together with any manufacturer's installation and maintenance instructions are handed over to the designated responsible person for the site premises.

It should then be made clear to all concerned that the commissioning process has been completed and responsibility for the plant is passed on to the appropriate personnel.

Warm air heating systems – 8

8 – Warm air heating systems

Principles of warm air heating

Free discharge systems – Indirect-fired

Indirect-fired gas heaters provide hot air into a room or workspace by the principle of convection.

'Convective heat transfer' relies on the movement of a gas (air) or fluid to carry heat and transfer that heat to other gases, fluids or solids.

The movement of the gas (air) or fluid can be natural, i.e. warm air rising due to temperature difference, or forced, i.e. fan impelled warm air movement.

The rate of heat transfer by convection between interfacing mediums will be increased if the velocity of the fluid or gas (air) is increased. However, there is an economical limit to increasing velocity based on the effective surface area of a heat exchanger giving up the heat to the transferring medium, gas or air, i.e. less heat will be given up to convection from a small heat exchanger surface.

The effective surface area of the heat exchanger can be increased by the use of fins or swirl plate mechanisms (see **Part 11 – Definitions**).

A gas-fired burner is used to heat a heat exchanger to a temperature of nominally 50°C. Once up to operating temperature, a fan operates to blow air across the surface of the heat exchanger before discharging through louvred outlet nozzles into the building to be heated.

A fan limit switch is provided to shut off the gas to the burner if the temperature of the heat exchanger rises to a level of nominally 90°C. The fan will continue to run to cool the heat exchanger after the burner has been extinguished, before switching off when a predetermined temperature is achieved.

When heating a work area using warm air systems, temperature distribution needs careful consideration (see Figure 8.1 for possible distribution arrangements).

The warm air emitted from the heater units needs to be free to discharge evenly over the occupied zone without obstructions, such as partition walls for instance. Where partitions are installed, the warm air should be distributed via a ducted outlet system.

However, when ducting warm air from one room to another, there may be certain situations where fire dampers or devices to restrict the spread of smoke will be required. Ducted systems are covered in more detail later in this Part.

A clear area will need to be left around the base of the heater units to allow the cooler air to re-circulate

Heat will generally be more evenly distributed if the system comprises of several individual, smaller heating units, rather than a single larger unit located centrally within the building.

However, the location of multiple heaters again needs careful consideration. If for instance, two heater units were located directly facing each other, the two warm air streams emitted from the heaters would collide forming a turbulent mass of air, whereby the warm air will tend to rise, being replaced by cold air (stratification).

Even without locating the heater units opposite each other, some stratification will take place due to the natural tendency for warm air to rise. In tall building areas particularly, such as distribution warehouses, the effects of stratification can result in temperature differences of 15°C to 20°C between the occupied space and the roof level (see Figure 8.2).

Stratification will result in wasted energy producing heat that rises to high level and not directly benefiting those people in the occupied zone of the building.

One solution to reduce the wasteful effects of stratification is to install suitable de-stratification fans at high-level to re-distribute the warmer air that would otherwise collect back down to the occupied zone (see Figure 8.3).

Figure 8.1 Possible suitable warm air distribution arrangements

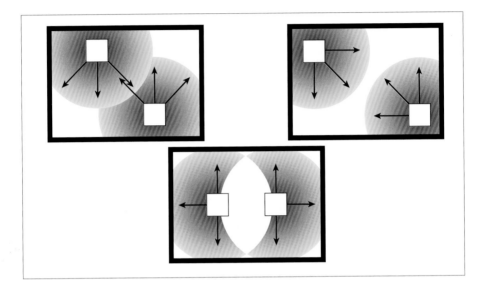

Figure 8.2 The effects of stratification between the occupied space and roof

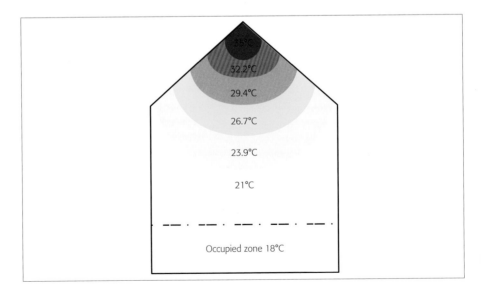

Figure 8.3 The effects of de-stratification fans between the occupied space and roof

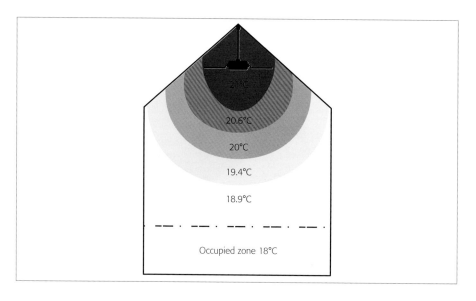

21°C

20.6°C

20°C

19.4°C

18.9°C

Occupied zone 18°C

De-stratification fans can be obtained for any particular working height, the higher they are mounted the more powerful the fan unit will need to be. Figure 8.4 illustrates the basic types available.

An alternative warm air heating system, which will reduce the effects of stratification and the need to install de-stratification fans is to provide the heating by means of a direct-fired system.

Free discharge systems – direct-fired

Direct gas-firing, as the name implies, takes air from outside which is then passed over a suitably sized gas burner and discharges the warm air, complete with the POC, directly into the space to be heated.

The obvious benefit is that without a heat exchanger and chimney/flue pipe there are no combustion losses to the outside atmosphere, which are associated with indirect firing heaters.

The only efficiency loss with direct firing is the latent heat of vaporisation of water created during the combustion process, which is approximately 8%. System efficiencies of a direct-fired heater can therefore be in the order of 92% based on gross thermal input whereas system efficiencies for indirect-fired heaters will be approximately 80% due to flue losses.

NG contains no sulphurous gases and the burner systems utilised for direct-fired heater units will ensure that, when correctly set, complete combustion produces only carbon dioxide (CO_2), water vapour (H_2O), nitrogen (N_2) and oxygen (O_2).

In view of the fact that these POC are being carried into the air, which will pass into the heated zone for occupants to breath, tight control of combustion is maintained so as to ensure the atmosphere is not contaminated.

Figure 8.4 Types of de-stratification fans

The European standard, BS EN 525: 2009 'Non-domestic direct gas-fired forced convection air heaters for space heating not exceeding a net heat input of 300kW' stipulates the limiting concentrations in the occupied zone for combustion gases attributable to the appliance as follows:

- CO_2 – 2500ppm

- CO – 10ppm

- NO (Nitrogen Oxide) – 5ppm

- NO_2 (Nitrogen Dioxide) – 1ppm

The HSE provide further guidance on exposure limits for various substances that can be encountered in the workplace via EH40/2005 'Workplace exposure limits'.

Workplace Exposure Limits (WELs) are provide for short term exposure (15 mins) and for long term exposure (8hr time weighted average (TWA), i.e. to cover a typical working shift), which should not be exceeded.

The WELs for POC are:

- CO – 30ppm (8hr TWA) & 200ppm (15 mins)

- CO_2 – 5000ppm (8hr TWA) & 15000ppm (15 mins)

With direct firing there is a constant stream of fresh air being provided through the heater unit. If this airflow is then designed to at least match the natural air change rate of the building, the contaminated atmosphere in the room is constantly being pushed out through small gaps present in the building structure, e.g. under doors, through window frames, through air bricks etc.

Figure 8.5 Principle of direct-fired heating

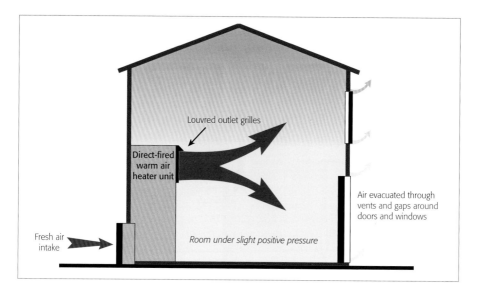

Louvred outlet grilles

Direct-fired warm air heater unit

Air evacuated through vents and gaps around doors and windows

Fresh air intake

Room under slight positive pressure

However, if the building is of modern insulated construction, it may be necessary to install additional exhaust vents to ensure the system works correctly. In some cases, these vents can be automatically modulated and linked to the heated air output from the unit.

In this way, the building will be maintained at a very slight positive pressure and the heated air percolates the complete building, minimising stratification, without the need to install additional air distribution fans or unsightly ductwork (see Figure 8.5).

Also helping to minimise stratification is the lower air temperature emitted from a direct-fired heater unit, which is typically about 40°C to 45°C. This is, again typically, about 15°C less than that from conventional indirect-fired heater units. However, some manufacturers build direct-fired systems with higher output temperatures.

Most direct-fired heater units incorporate modulating burner systems, which can have turndown ratios of up to 25:1, to provide accurate control of heat input whilst maintaining a constant supply of fresh air to ventilate the building.

With the burner automatically switched off when the room thermostat is satisfied, the heater unit can also be used to provide ventilation in the summer period.

The design of a direct-fired system is usually undertaken by manufacturers of the heater to ensure correct application of the system.

The system technique has advantages where heated air needs to be supplied as part of the building environment control, where the building's atmosphere is contaminated by industrial processes for example.

Figure 8.6 Simple overhead duct system to a single storey building

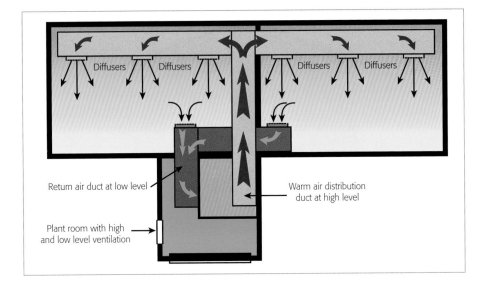

They are commonly found in commercial kitchens; swimming pools; industrial machine shops; welding/fabrication workshops and also churches (due to low frequency of operation and high heat losses).

Ducted air distribution

Where the building structure makes it impossible or difficult to achieve an even warm air distribution using free discharge units, whether direct or indirect-fired, or if noise levels are a consideration, then it may prove beneficial to install a ducted warm air distribution system; the heater unit(s) can be located away from the room/areas to be heated, perhaps in a separate plant room.

The duct system will be specifically designed for each application to distribute the correct amount of warm air to individual areas or rooms by varying duct size and diffuser settings to any particular area.

Where distributing ductwork passes through a building from room to room, it will be necessary to consider building regulations and/or fire regulations.

This may necessitate provision of smoke detectors linked to fire dampers within the duct system. In the event of a fire, the spread of smoke and fumes from one area to another, through the duct system, will be prevented.

Figure 8.6 illustrates a plan arrangement of a simple overhead duct system to a single storey building.

Insulated warm air distribution ducts will normally be located at a high level in the building, perhaps under the ceiling or above a false ceiling, with stub ducts into the area to be heated. Each stub duct is fitted with a diffuser and balancing damper to direct the correct amount of warm air into the room.

Figure 8.7 Typical floor standing type warm air heater unit (indirect-fired)

Air velocities of 3.5 metres per second (m/s) to 5m/s, dependent on room height, are likely to be needed to distribute the warm air down to floor level.

A separate duct system is installed at low level to ensure a path of return air to the inlet of the heater unit.

Where any particular building is constructed with various sized rooms and/or different occupational requirements e.g. occupied times, sedentary (requiring a sitting position) or manual work, with a wide variation of heat demand, the use of multiple heater units can prove to be more efficient.

Typical warm air heater units

Free discharge systems – floor standing

These units are designed to draw in air at low level using a fan to blow the air over a heat exchanger and then discharge the heated air through nozzles. The nozzles can be swivelled 360° about their axis to permit free blowing of warm air in any direction (see Figure 8.7).

The stainless steel combustion chamber and heat exchanger are housed within a double skin cabinet, which will usually be insulated or air cooled to prevent heat loss and reduce outer casing temperature (see Figure 8.8).

These units are commonly available in heat output capacities between 30kW and 800kW, weighing between 230kg and 3800kg operating on NG and LPG. The fan motor to larger units will usually require a 3 phase 415V electrical supply.

Figure 8.8 Cut away section of an indirect-fired floor standing warm air heater

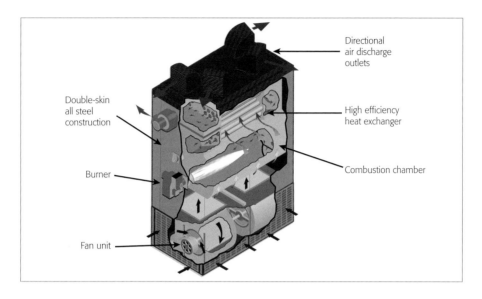

A combined fan and limit switch is fitted, which will delay the air distribution fan start until the heat exchanger has reached the normal operating temperature, nominally 50°C. On shutdown of the burner, the fan will continue to run to cool the heat exchanger down to nominally 30°C. In the event of overheating, the limit switch will turn the burner off at nominally 90°C – 100°C and will require manual resetting.

The gas burner system fitted to the heater units will normally be of the forced draught type incorporating full light-up sequence control with either rectification or ultraviolet cell flame supervision.

However, some smaller units (up to nominally 100kW) are available with atmospheric burners and thermo-electric flame supervision device.

Alternatively, where the application is appropriate (as described in **Principles of warm air heating**), a direct-fired air heater can offer certain benefits in terms of efficiency and internal environment.

The freestanding, direct-fired heater will operate slightly differently to that of the indirect-fired heater.

When the unit is switched on (via time-switch and thermostat) the main air fan will start immediately as this fan will be providing the through flow of air for combustion. After the appropriate air purge time has elapsed, the gas burner will be ignited and the heat generated at the burner transferred directly to the air passing through into the work space.

The air fan will then run continuously to provide a constant supply of fresh air; the burner will then modulate to match the heat demand of the room to be heated.

Figure 8.9 Cut away section of a direct-fired floor standing warm air heater

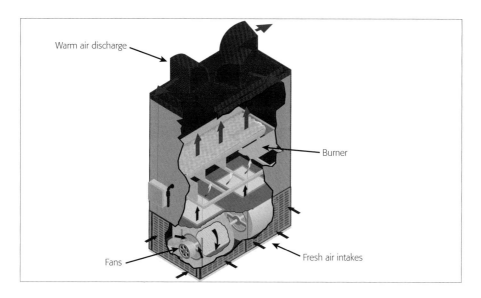

The weight of these standard type units will be less than the indirect-fired equivalent, due to the absence of the heat exchanger (see Figure 8.9).

Free discharge systems – suspended

These units are designed to draw in air at the rear of the heater using a fan to blow the air over a heat exchanger and then discharge the heated air through horizontal or vertical louvres at the front of the unit (see Figure 8.10).

The standard versions of these units normally have aluminised steel heat exchangers, although stainless steel heat exchangers are generally available as an option if the heater is to be located in contaminated atmospheres (see Figure 8.11).

These units are commonly available in heat output capacities between 12kW and 150kW, weighing between 40kg and 400kg for the standard free blowing heaters and can operate on NG or LPG.

Fan and limit switches are fitted, which will delay the air distribution fan start until the heat exchanger has reached the normal operating temperature, nominally 50°C. On shutdown of the burner, the fan will continue to run to cool the heat exchanger down to nominally 30°C. In the event of overheating, the limit switch will turn the burner off at nominally 90°C – 100°C and will require manual resetting.

The gas burner system fitted to the heater units will depend on the particular manufacturer's model, but will be either:

*   open-flued unit – with natural draught bar burners incorporating thermo-electric flame supervision and multi-functional control valve; or

*   room-sealed unit – fan assisted combustion through a tube heat exchanger with precisely matched injector incorporating electronic sequence controller and rectification flame supervision.

Figure 8.10 Typical examples of suspended unit air heater

Figure 8.11 Cut away section of a typical suspended type open-flued warm air heater

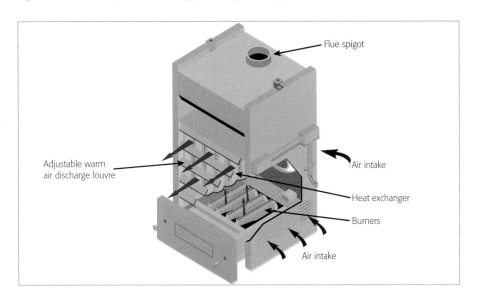

Figure 8.12 Propeller fan heater

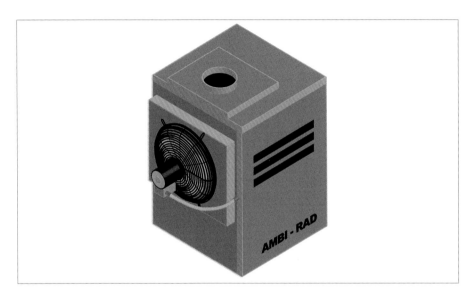

Suspended unit air heaters – propeller (axial) fan

Propeller fan units are provided as the basic heater unit to provide good air distribution with low noise levels. For effective warm air distribution down to floor-level the mounting height will be nominally 2.5m to 3m to the underside of the unit (see Figure 8.12).

Suspended unit air heaters – centrifugal fan

Belt driven centrifugal fan units are used where higher airflows are required allowing the units to be fitted at greater mounting heights than the standard model, up to 5m to the underside on certain models (see Figure 8.13).

A range of outlets are available to direct the warm air where it's needed, for example 30°, 60°, 90° downturn nozzles or down-flow plenum where a vertical discharge is required (see Figures 8.14, 8.15 and 8.16).

Suspended unit air heaters – fan compartment

The fan compartment unit is designed to meet a wider range of air handling applications. The units are again fitted with the centrifugal fan arrangements, but the cabinet will facilitate installation of additional options, such as filter units and connection of return air ductwork (see Figure 8.17).

Although these cabinet arrangements can be suspended it might be more appropriate in most cases to mount them on concrete plinths or steelwork from the floor, due to the additional weight of connecting ductwork etc.

Extra options are available to optimise performance of the heater system such as modulating burner control; airflow proving, dirty filter sensing and constant fan running for combined heating/ventilating applications where fresh air is required.

Figure 8.13 Centrifugal fan heater

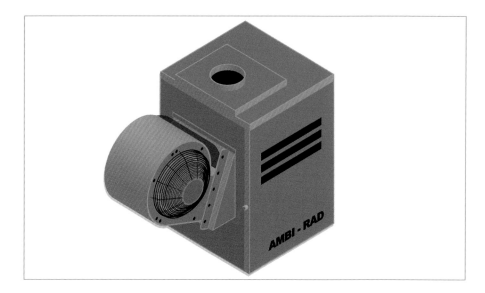

Figure 8.14 Example of an outlet extension duct – downflow plenum

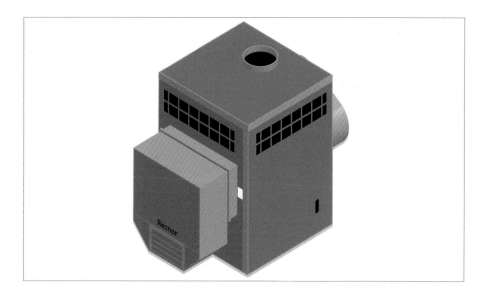

Figure 8.15 Example of an outlet extension duct – two-way discharge

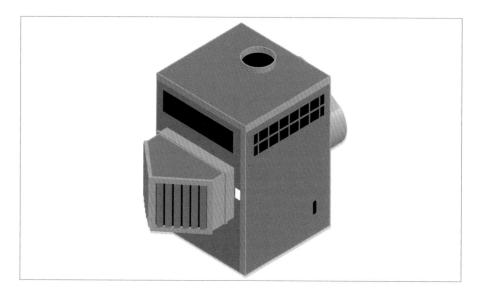

Figure 8.16 Example of an outlet extension duct – downturn nozzle

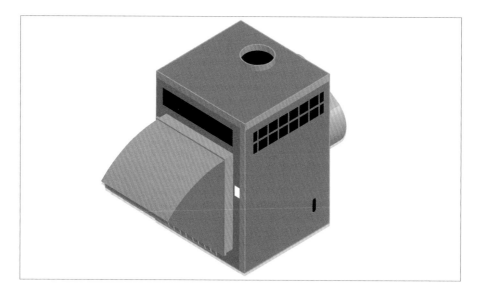

Figure 8.17 Fan compartment heater unit

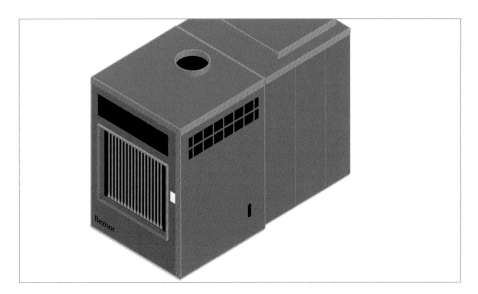

Where units are to provide fresh air for ventilation, it will usually be preferable to use a heater unit with stainless steel heat exchanger to minimise possibility of corrosion from atmospheric conditions.

For units connecting to return air duct systems, a damper facility can be incorporated in the cabinet to control the re-circulated air either manually or automatically. Automatic operation can be achieved using two, three position or fully modulating damper actuators.

Condensing unit air heaters

Part L of the schedule to the Building Regulations (England and Wales) is concerned with the conservation of fuel and power in buildings.

The efficiency of heating systems fitted in buildings used for commercial and industrial purposes, are covered by Approved Documents L2A (ADL2A) 'Conservation of fuel and power in new buildings other than dwellings' and L2B (ADL2B) 'Conservation of fuel and power in existing buildings other than dwellings'.

Note: For Northern Ireland refer to 'Technical Booklet F2 2012' and for Scotland 'Technical handbook non-domestic 2011'.

ADL2A, ADL2B & 'Non-Domestic Building Services Compliance Guide' provide guidance on how to satisfy the energy performance requirements in non-domestic buildings. The Building Regulations are covered in greater detail in **Part 2 – Gas and associated legislation.**

Figure 8.18 High efficiency condensing unit air heater

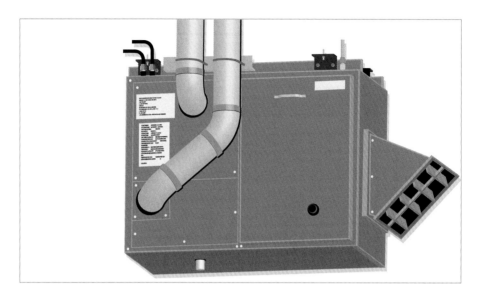

To improve energy performance of heating systems in non-domestic buildings, it is possible to fit suspended unit air heaters that can operate in condensing mode (see Figure 8.18).

These appliances are room-sealed (Type C classification) in accordance with PD CEN/TR 1749: 2009 'European scheme for the classification of gas appliances according to the method of evacuation of the combustion products (types)' – appliances with atmospheric burners firing into a conventional heat exchanger.

The POC generated at the burner, rise through the primary heat exchanger and then pass through an additional 'secondary' heat exchanger. An exhaust fan is fitted at the outlet of the secondary heat exchanger to ventilate the combustion chamber and pressurise the flue.

By passing the POC through this secondary heat exchanger, the temperature of the gases will drop below their dew point (55°C nominally), such that the latent heat is extracted from the water vapour. In this way, a heater efficiency of approximately 92% based on gross calorific value (CV) is achieved.

The operating principle can be seen in the Figure 8.19

The amount of condensate produced is approximately 1 litre per hour for every 20kW of gross thermal input. This condensate will be mildly acidic with a Potential of Hydrogen (pH) value up to 3.0 and therefore, will need to be evacuated from the heater using suitable sanitary plumbing material i.e. not copper tube or copper based alloy.

Figure 8.19 Operating principle of condensing air heater

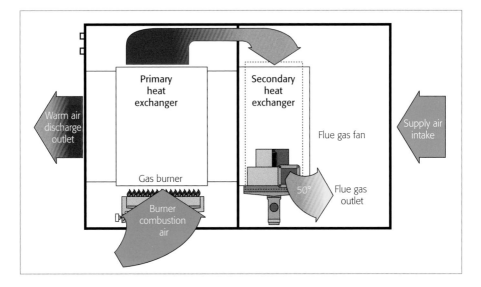

Ducted systems

Ductwork distribution systems can be utilised to distribute warm air evenly through a room or throughout a building, if the building structure is suitable (see **Principles of warm air heating** in this Part).

The ductwork can be connected to most of the free discharge type heater units, floor standing or suspended, provided the air distribution fan is designed correctly to overcome the pressure resistance of the outlet duct system (see Figure 8.20).

However, the propeller (axial) fanned heater units would not usually be appropriate for this application. The axial fan on these heaters will not generate sufficient air movement to overcome duct resistance and will cause the heat exchanger to overheat leading to premature heater failure.

Make up air – direct-fired systems

There are many commercial and industrial premises where fumes are generated from various processes, to warrant the removal of these fumes by air extraction systems, e.g. restaurant kitchens; vehicle repair garages; machine process factories etc.

When this air is removed by extraction, it becomes necessary to 'make-up' the lost air with that of fresh clean air to continually replenish the atmosphere in the room.

Where incoming fresh cold air is replacing the contaminated warm air, the incoming replacement air will need to be heated directly. Often the systems will utilise ductwork to distribute the clean air more effectively.

Figure 8.20 Typical fanned compartment heater with air inlet and distribution duct

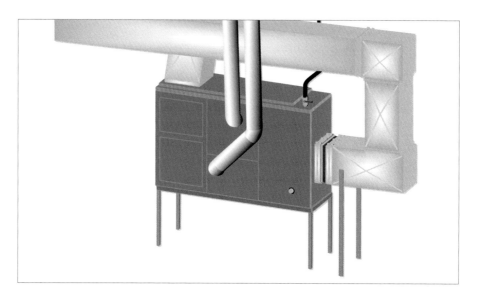

The heater section of the units will comprise the features of normal direct-fired heaters, that of a fan downstream of a gas burner drawing fresh air in over the burner to distribute into the occupied space.

The make-up air heater can be built to match system requirements in the same way as the suspended type cabinet heaters, i.e. by providing filtration, air recirculation ducts and automatic dampers to control the amount of recirculation (see Figure 8.21).

In addition to the heater section of the unit, cooling coils can be fitted for air conditioning in summer months.

The make-up air unit will be fabricated specifically to meet the requirements of each particular application. But standard heater modules, which are at the heart of the system, are generally available from 30kW up to 900k.

Figure 8.21 Direct-fired make-up air unit

Installation requirements

The following information provided in this Part offers general guidance notes for the installation of the types of heaters previously described. Specific information relating to the installation of any particular heater or system will be described in the appropriate manufacturer's instructions.

Particular installation normative reference documents include:

- BS 6230: 2011 'Specification for installation of gas-fired forced convection air heaters for commercial and industrial space heating (2nd and 3rd family gases)

- IGEM/UP/2 (Edition 2) 'Installation pipework on industrial and commercial premises'

- IGE/UP/10 (Edition 3) with amendments October 2010 'Installation of flued gas appliances in industrial and commercial premises incorporating specific requirements for appliances fired by bio-fuels'

- BS 5440-1: 2008 Flueing and ventilation for gas appliances of rated input not exceeding 70kW net (1st, 2nd and 3rd family gases) – Part 1: Specification for installation for gas appliances to chimneys and for maintenance of chimneys

- IGEM/UP/18 'Gas installations for vehicle repair and body shops'

- EH40/2005 Workplace Exposure Limits Containing the list of workplace exposure limits for use with the Control of Substances Hazardous to Health Regulations (as amended) – available freely as a pdf download from www.hse.gov.uk/pubns/books/eh40.htm or as a hard copy from HSE Books

In order to assist gas operatives, and others concerned with the installation of gas appliances in to non-domestic premises, the Communities and Local Government (CLG) produce a free guide, entitled 'Non-Domestic Building Services Compliance Guide', which provides the reader guidance on types of heating systems and relative efficiencies required when installing heating in new or existing premises.

Note: The guide can be obtained by visiting http://www.planningportal.gov.uk/buildingre gulations/approveddocuments/partl/bcassoc iateddocuments9/compliance

For further information regarding the Building Regulations see **Part 2 – Gas and associated legislation – Building Regulations (England and Wales)** in this manual.

When designing any warm air heating system there are many areas of consideration; collaboration will be essential between those involved such as the gas operative, the occupier, and the building owner.

There may also be the need to inform local authorities for appropriate planning consent (e.g. for listed buildings), fire authorities, building insurers etc.

Additionally, the installation will need to take in to account the requirements of Building Regulations applicable to the geographical area concerned (see **Part 2 – Gas and associated legislation** for further guidance).

Matters to be considered include:

*   heater manufacturer's installation requirements

*   minimum appliance efficiency (including associated controls) as required by Building Regulations

*   building construction (fabric heat losses, suitability and proximity of materials etc.)

*   flueing requirements

*   ventilation requirements

*   gas availability (type, site capacity and pressure)

*   electricity requirements

*   site access and liaison with associated trades, etc.

Note: The above list should not be regarded as exhaustive.

Competence

Work carried out relative to the installation of warm air heating systems, needs to be undertaken by operatives who are experienced in such work and who hold the relevant certificates of competence, obtained through the ACS and/or aligned N/SVQ's.

Note: For more information regarding commercial ACS categories refer to Part 3 – 'Competency' and Essential Gas Safety – Non-Domestic manual (Order Ref: ND1), see Part 13 – CORGI*direct* Publications.

Appliance suitability

Any appliance or system must be fit for the purpose for which it is intended. This can be ascertained from a variety of sources:

* CE mark – ensure the appliance carries a CE mark together with the identification number of the notified body responsible for essential information – maximum input gas pressure, burner gas pressure, kW heat input rating etc.

* the aforementioned data plate should also display;

    - country of import (e.g. GB)

    - type of gas and supply pressure (e.g. G20 = natural gas, G30 = butane, G31 = propane)

    - heater/multi burner system category as defined in BS EN 437: 2003 + A1: 2009 'Test gases. Test pressures. Appliance categories' (e.g. I_{2H}, I_{2H3P} etc.)

General location requirements

Heaters will need to be sited clear of combustible materials e.g. fabric of the building, particularly with some older structured buildings where roof support beams may be of timber construction. Manufacturers will normally provide guidance on minimum distances from combustible material.

Appliance manufacturer's instructions will also provide clearance requirements for adequate maintenance access.

All items that are liable to wear or otherwise require regular attention such as burners, fans and control devices will need adequate access provision to allow withdrawal/removal without causing damage to other components on the unit or system.

In factory and warehouse buildings there is likely to be significant movement of fork lift trucks, or other hoisting mechanisms. Any heater units will need to be located with due consideration of such equipment to avoid any possibility of heater damage.

Furthermore, where there is a perceived risk of physical damage to the heater units, it may be necessary to provide some form of mechanical barrier in order to minimise risk.

If the heater is to be located in atmospheres containing vapours, gases or airborne dust, the heater unit will need to be located adjacent to external walls, to allow fresh air to be ducted to the burner for combustion.

Condensing air heaters will involve additional considerations particularly regarding the condensate removal.

The condensate disposal pipe will need to run indoors to avoid the possibility of freezing in winter. The location of the heater can sometimes be dictated by the nearest access to soak away drainage facilities.

Also, condensing heaters can have a tendency to form a plume of water vapour from the flue terminal; consideration will therefore be required as to the effect of this pluming on surrounding building structure, materials and neighbouring buildings.

Restricted locations

Individual heaters and associated air distribution systems will need to be installed in locations that are deemed suitable. In industry particularly, there will be a wide variety of applications and processes.

Therefore, due consideration needs to be given to the likely environmental conditions:

* corrosive or salt-laden atmospheres will obviously have an affect on metallic components, particularly burners and controls, e.g. metal plating/treatment workshops

* dusts and vapours given off from plastic forming processes. Cleaning or curing applications when passed into the atmosphere will be drawn into the combustion chamber and may subsequently produce harmful gases

* chlorine laden atmospheres (e.g. at swimming pools/baths) – when atmosphere is laden with chlorine in the vicinity of a permanent heat source, such as permanent gas pilot light, the chlorine will break down to base component chemicals, including acids which will attack the metallic components of the heater and corrode very rapidly.

* high velocity air movement in the vicinity of the burner will affect its performance and may prevent complete combustion from taking place

* petroleum or heavier than air flammable vapours represents a significant risk of explosion. Therefore, heaters or multi-burner systems must not be installed in areas that are classified as being hazardous in accordance with BS EN 60079-10-1: 2009 'Explosive atmospheres. Classification of areas. Explosive gas atmospheres'.

Where the area is not classified as being hazardous but heavier than air vapours might be present – as a result of spillage at a vehicle workshop/garage for example – any gas heater located in such an area, will need to be sited so as not to represent an ignition source to the fuel spillage i.e. the base of the heater will need to be mounted at least 1.8m above floor level.

In addition, electricity supply cables and switchgear will need to be sited above 1.2m or appropriately protected as defined in BS EN 60079-10-1.

Note: For additional guidance for garages, refer to IGEM/UP/18 'Gas installations for vehicle repair and body shops'

BS EN 60079-10-1 defines areas in distinct zones, i.e. zone 0, 1 and 2, relative to known hazards that may be present. Increasing the amount of ventilation in a hazardous area may in certain cases change the zone rating of an area and may then permit installation of gas-fired heater units.

* LPG fuelled appliances with automatic ignition or permanent pilot light cannot be located in a totally enclosed room below ground e.g. cellar or basement.

If there is any doubt as to the suitability of any heating system for a given application, always seek advice from the appliance manufacturer(s).

Most manufacturers will prefer to give, often free advice, in order to ensure the heating is appropriate for the application, rather than risk the possibility of inheriting bad publicity resulting from unsuitable/poor installations.

Air supply/ventilation

Regulation 26(1) of the GSIUR states –

"No person shall install a gas appliance unless it can be used without constituting a danger to any person".

To achieve complete combustion it is essential to ensure that the appliance is supplied with an adequate supply of clean fresh air (ventilation). This will take into account the type of flue fitted to the appliance and any other appliances in the same area, along with any cooling requirements for the appliance and the occupants of the workspace in which the appliance is sited.

Note: For information relating to general ventilation requirements in non-domestic establishments, see Essential Gas Safety – Non-Domestic (Order Ref: ND1), see Part 13 – CORGI*direct* Publications.

In addition to manufacturer's instructions, the industry standards referred to for ventilation installations relative to warm air heating systems are as follows:

* BS 6230

* IGE/UP/10 (Edition 3)

* BS 5440-2: 2009 'Flueing and ventilation for gas appliances of rated input not exceeding 70kW net (1st, 2nd and 3rd family gases) – Part 2: Specification for the installation and maintenance of ventilation provision for gas appliances'.

Appliance support/mounting

Floor mounted heaters are usually self-supporting due to the weight of the unit, but may sometimes require the extra stability of anchor bolts to the floor.

Individual unit heaters can be suspended from the building support steelwork – see Figure 8.22, or suspended from appropriate brackets mounted from a suitable wall structure – see Figure 8.23.

The location of a particular heater can often be dictated by the need for adequate support or suspension facility. The structure of the building may restrict possible locations for this facility.

Any support or suspension facility will need to be of sufficient number and strength to support the full weight of the heater and its associated system (outlet duct). Also the particular means of support will need to be constructed of suitable materials not liable to corrosion.

The building structure will need to be sufficiently strong to withstand the extra load placed upon it by the heaters, or heater system, complete with the associated chimney/flue pipe system.

In the event of suitable roof steelwork not being available, additional steelwork should be fitted to enable hangers to be used for suspending the heaters.

If there are any doubts as to the strength or suitability of building structure to which heaters or system is to be suspended, reference will need to be made to the building consultant/architect/structural engineer.

The hanging attachments to overhead steelwork etc. need to be purpose made to good sound engineering practice or of a proprietary type fixing.

Figure 8.22 Typical suspension detail

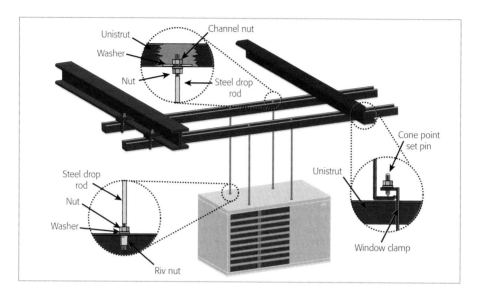

Figure 8.23 Typical bracket support detail

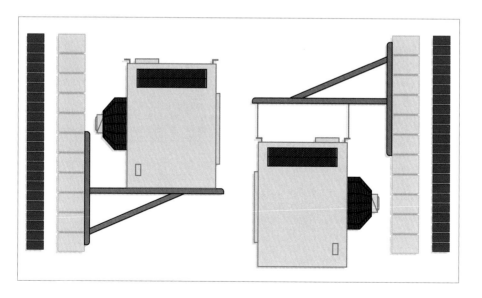

Gas installation

The GSIUR require gas fitting operatives to ensure that gas installation pipework and fittings are installed safely with due regard to the location of other services, e.g. other pipes; pipe supports; drains; sewers; cables; conduits and electrical control equipment.

Gas operatives will also need to be aware of the limitations of the building structure when installing gas equipment and pipework.

Note: For information relating to the general requirements for gas pipework in non-domestic establishments, see Essential Gas Safety – Non-Domestic (Order Ref: ND1), see Part 13 – CORGI*direct* Publications.

The industry standard referred to generally for non-domestic pipework installations is IGEM/UP/2 (Edition 2) 'Installation pipework on industrial and commercial premises.

Gas connections

Regulation 26(2) of the GSIUR states –

"No person shall connect a flued domestic gas appliance to the gas supply system except by a permanently fixed rigid pipe".

However, Section 13 of IGEM/UP/2 (Edition 2) states that –

"The use of a flexible connection shall be considered in situations where it is known, or anticipated, that pipework will be subject to vibration, movement, expansion or strain".

It is reasonable to expect that some individual suspended unit air heaters will be subjected to a certain amount of movement, particularly when suspended by drop rods that could be 2 or 3 metres long for example. Guidance then should be sought from the appliance manufacturer as to the need for gas flexible connection.

Where manufacturers stipulate that the final gas connection to the burner/heater unit is by means of a flexible connection, that connection needs to be constructed out of suitable material – stainless steel for example, complying with BS 6501-1: 2004 'Metal hose assemblies. Guidance on the construction and use of corrugated hose assemblies' and/or BS EN ISO 10380: 2003 'Pipework. Corrugated metal hoses and hose assemblies'.

For added protection, it is advisable that the flexible connection incorporates a surface cover. This may be a plastic sleeve, or if there is risk of physical damage, over-braided stainless steel.

Note: The decision as to the appropriate sleeving material to use will need to be made, bearing in mind the environment in which the heater and therefore, its gas connection will be exposed to. For example, BS 6501 strongly recommends that plastic coatings containing either sulphur or chlorine should not be used.

Alternatively, it is acceptable to use a gas 'Caterflex' which is a plastic coated stainless steel flexible pipe manufactured to BS 669-2: 1997 'Flexible hoses, end fittings and sockets for gas burning appliances. Specification for corrugated metallic flexible hoses, covers, end fittings and sockets for catering appliances burning 1st, 2nd and 3rd family gases' and incorporates a self-sealing plug and socket.

Any gas flexible pipe needs to be installed such that there are no sharp 90° bends and the flex is not subjected to stress or torsion.

A slow radius bend of 180° onto the burner will usually be required, to take up expansion and contraction created at the heater by its heating and cooling cycles.

Where a self-sealing plug and socket is not used, a union joint will be required at one end of the flexible pipe to facilitate disconnection of the gas pipe for maintenance of the burner/heater unit. Installation should ensure that the flexible does not twist when the union is tightened.

In any event, a 90° operable gas isolation valve will need to be installed immediately upstream of the flexible pipe assembly.

Due consideration should also be given to the available gas supply pressure in the pipework system feeding the heater.

On some industrial and commercial sites it may be possible for the metered supply to have elevated gas pressure, or on large sites, such wide variation on load characteristics as to create significant variation in the available gas pressure.

Where either of these situations exist, it will be necessary to check with the appliance manufacturers to establish if there is a need for fitting an additional gas regulator upstream of the appliance control system.

In certain cases, particularly where the supply pressure to the appliance is 75mbar or more, there will need to be adequate automatic means for preventing the appliance and associated pipework from being subjected to this higher pressure e.g. in the event of regulator failure. This 'protection' would usually be by means of a regulator with integral overpressure slam-shut valve incorporated.

Electrical connections

Non domestic properties normally utilise three-phase electrical systems due to the high current demands of the electrical equipment installed and used.

When gas operatives are required to work on gas appliance/equipment, which utilise either three-phase or single-phase electricity, it is important that those gas operatives are competent and have the relevant knowledge and experience to work safely on those systems and be able to carry out a thorough risk assessment of any potential hazards that may be involved.

For further information on risk assessments see Essential Gas Safety – Non-Domestic – Part 3 – Risk assessments.

Important: The Electricity at Work Regulations 1989 (EWR) apply to any operative carrying out any type of electrical work. It is a requirement of the EWR that the person is competent to undertake the work.

Where any doubt exists with regards to competency or lack of knowledge/experience to work on those electrical systems, or where a risk assessment identifies that the hazard would be unacceptable, no work should be undertaken.

Working on three-phase equipment, supplies and four wire supplies

In the UK the 400V, 4-wire three-phase and neutral (TPN) method is widely used for the distribution of supplies within commercial and small industrial installations.

For example, on large wet central heating systems the pump may operate at 400V from a three-phase supply, while other components such as thermostats and time switches may be 230V single-phase or sometimes even less, for example 24V, which is classed as Extra Low Voltage (ELV).

On commercial warm air heaters, the fans (combustion and/or circulating) may be three-phase, while once again other components may be single-phase.

It must be remembered that each of these phases is carrying 230V. A balanced three-phase circuit does not require a neutral for it to be able to operate. This is because there is always a return path for the current through one of the other two phases.

There are a number of signs to look for when trying to identify the voltage on non-domestic gas installations, two most basic signs are:

1. Appliance data plate – one of the simplest methods of determining the voltage to a non-domestic gas appliance is by checking the information on the appliance data plate, assuming of course a plate is present.

   The information displayed will typically consist of:

   - supply voltage – 400V

   - supply frequency – 50Hz

   - no of phases – 3

   - no of wires – 4

   - power consumption – 4.0kW

   - fuse rating – 400V, 25A HBC

   - fuse rating – 230V, 5A

2. Installation wiring – all installation wiring falls within the scope of BS 7671: 2008 (Incorporating amendment No. 1) 2011: 'Requirements for electrical installations. IET Wiring Regulations Seventeenth Edition' and therefore requires completion of electrical certification when installed.

   The existence of a three-phase supply can often be confirmed by the wiring colour code, this will need to be tested to ensure it is correctly sized etc. before undertaking any electrical installation work.

Remember, all electrical installations should be in accordance with BS 7671 and the supplied manufacturer's wiring guides, especially for the correct connection of supplies.

Of equal importance is the correct routing of cables, including suitable supports where required and the correct securing of cables to individual pieces of equipment.

Each appliance and its associated controls needs to have its own means of electrical isolation, which should be situated in close proximity to the equipment it serves and have suitable means for the application of securing devices, i.e. hasps or padlocks for ensuring that the appliance cannot be re-energised, whilst being worked on by an operative (this should be highlighted as part of a risk assessment).

Chimney systems

Appliances and their flues are classified by a 'European Committee for Standardization' document entitled 'PD CEN/TR 1749: 2009 'European scheme for the classification of gas appliances according to the method of evacuation of the products of combustion (types)'.

The purpose of this document is to harmonise, across Europe, the classification of appliances burning combustible gases.

PD CEN/TR 1749 separates gas appliances into three types:

- Type A – flueless

- Type B – open-flued (including systems formerly known as 'closed' flue)

- Type C – room-sealed.

For further detail on the correct flue classification for general appliance types, see Essential Gas Safety – Non-Domestic (Order Ref: ND1, see **Part 13 – CORGI***direct* **Publications**).

Change to terminology

European standardization has also brought about changes to terminology typically used within the gas industry. One such change is with regards to what we in the UK have referred to as 'flues'.

Under European terminology 'flue' describes the passage of the POC and not the material or structure for the transportation of that POC. Where we in the UK have used the term flue to also describe the material/structure, this is defined as either 'chimney' or 'chimney component'.

Therefore and in the main, this manual will use the newer term (chimney) when describing flue construction/material as well the term 'flue pipe' (connecting and appliance to a chimney), as appropriate and 'flue' (when describing the passage of POC).

Note: Current Standards use a mixture of terms, including the older term of flue and therefore, where used will still be relevant to that Standard.

Chimney installation

Any chimney/flue pipe system will need to be installed in accordance with manufacturer's instructions to ensure safe and complete evacuation of combustion products to the atmosphere.

Chimney components and their materials of construction will need to be in accordance with the appropriate British Standards.

Note: For details on the general appliance chimney/flue pipe installation, see Essential Gas Safety – Non-Domestic (Order Ref: ND1, see Part 13 – CORGIdirect Publications).

Single heater installation – Type B

Unless otherwise stated in manufacturer's instructions, it will be necessary to install the chimney/flue pipe system so that it is supported independently of the heater unit.

The manufacturer's chimney adaptor piece should always be used to ensure correct connection to the flue spigot and a complete seal is obtained. Consider also, that provision will need to be made so that the chimney/flue pipe system can be disconnected from the heater to allow inspection and servicing.

The flue terminal will need to be positioned so as to ensure that no POC are allowed to re-enter the building or re-circulate into the combustion air intake.

If the terminal is located less than 2m above ground level (e.g. wall termination), then it needs to be fitted with a suitable guard to prevent possible blockage or injury to persons.

For air heaters rated below 70kW net, the flue termination will need to conform to the requirements of BS 5440-1.

Note: It is generally recommended that consideration be given to the fitting of a suitable terminal where the flue terminates in free atmosphere. However, where the flue diameter is less than 170mm an approved terminal will be required.

Where the route of the chimney/flue pipe is such that horizontal runs are unavoidable, the appliance manufacturers should be consulted regarding maximum horizontal lengths.

With natural draught appliances, it is usually possible to purchase an exhaust fan from the manufacturers for the particular heater to be installed, which will assist in removing the POC.

Where any such fan is utilised, it will need to be installed in accordance with the manufacturer's instruction and interlocked to the gas supply in accordance with the GSIUR, to ensure that the gas burner will be shutdown in the event of fan failure.

Prevention of condensation within the flue is an important design factor, manufacturers will usually recommend the use of an approved double wall chimney/flue pipe to minimise the likelihood of this occurring.

Where condensation in the flue is unavoidable, appropriate provision will need to be made for the condensate to flow freely to a disposal point.

Current standards require that a data plate is displayed on existing chimney/flue pipe systems detailing the various performance characteristics.

Where a heater is to be connected to such an existing chimney/flue pipe system, the gas operative will need to verify with the heater manufacturers that the information displayed on the chimney/flue pipe data plate is appropriate for the connection of the proposed heater(s).

Double heater installation – Type B

Common chimney assemblies can only be installed where specifically recommended by the appliance manufacturer.

Certain manufacturers produce twin heater units, whereby burner systems are duplicated on each side of the heater unit. It will be appropriate in these circumstances to install the common chimney/flue pipe header unit supplied by the manufacturers (see Figure 8.24).

Single heater installation – Type C

Any room-sealed flue system will need to be assembled and installed with manufacturer's approved components.

When using the concentric termination illustrated in Figure 8.25, then the complete system will need to be installed with manufacturer's approved components and in accordance with the manufacturer's instruction.

Note: This type of system is regarded as an integral part of the appliance and formed the basis on which the appliance was tested to achieve its CE approval. To deviate from this would therefore contravene the GSIUR.

The flue terminal will need to be positioned so as to ensure that no POC are allowed to re-enter the building or re-circulate into the combustion air intake.

If the terminal is located less than 2m above ground level, then it needs to be fitted with a suitable guard to prevent possible blockage or injury to persons.

Figure 8.24 Twin heater unit showing common chimney/flue pipe arrangement

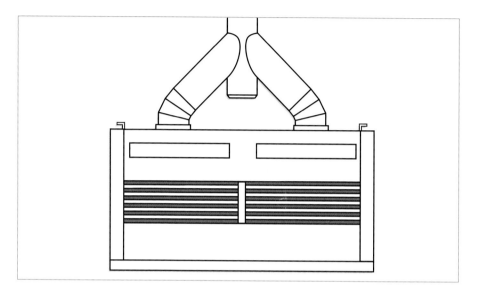

Figure 8.25 Horizontal and vertical room-sealed flue system

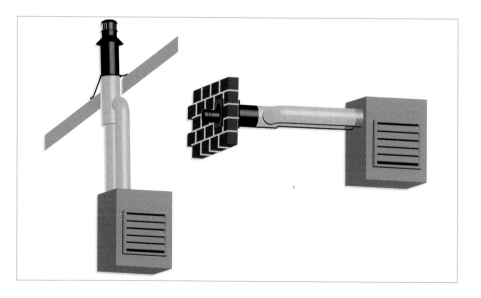

Commissioning

Manufacturer's instructions will provide specific information for commissioning any particular appliance and its associated equipment.

In addition, IGE/UP/4 (Edition 2) 'Commissioning of gas-fired plant on industrial and commercial premises', offers further generic guidance for the commissioning process.

Correct commissioning of an appliance in accordance with the manufacturer's instructions is as critical to a safe and efficient installation, as its initial installation.

Therefore, where the installation can not be commissioned immediately after installation, it should be isolated from the gas supply and suitable notification attached (this fact also needs to be documented on any appropriate paperwork) to advise the reader that the installation is un-commissioned.

The following is provided for information purposes only; applying to indirect-fired air heaters and should not be regarded as substitute for the source documents.

Planning

Before travelling to site, a certain amount of preparation will be necessary, including:

- ensure all relevant manufacturer's information is available

- all relevant drawings and system plans are available

- all gas tightness test/purge certificates have been completed by the upstream pipework installer

- all electrical test certificates have been completed by the electrical installer

- risk analysis has been completed

- COSHH statements prepared, where relevant

- any 'hot work' or 'permit to work' has been granted (see Essential Gas Safety – Non-domestic – Part 3 – Risk assessments (Order Ref: ND1), see **Part 13 – CORGI*direct* Publications**

- access facilities are available. Due to the fixing height of suspended heating systems, suitable scaffolding or hydraulic lift facilities (cherry pickers) will usually be required

- all necessary tools are available and suitable for the purpose and are of sound quality. Any electronic instruments need to be correctly calibrated and certificated accordingly

- a written plan of required work is completed (usually provided in manufacturer's instructions).

Inspection

Once on site, a complete inspection of the installation will need to be carried out to ensure that:

- all gas and electrical supplies are suitably sized, correctly located and supported and that the installation has been installed in accordance with the manufacturer's instructions and relevant standards

- assembly is complete and all components are fit for the intended purpose

- the appliance has been correctly supported/suspended; it is level as required by manufacturer's instructions and is stable with adequate clearance for maintenance and from combustible materials

- gas and electrical supplies have been isolated.

Activation – dry run

Subject to satisfactory inspection, the next phase of the commissioning process is to carry out preliminary physical checks with fuel isolated:

*   a suitable gas tightness test is undertaken on the appliance gas connection and its isolation valve(s)

Note: For detailed guidance on the correct procedures for testing gas pipework in non-domestic establishments, see Essential Gas Safety – Non-Domestic (Order Ref: ND1), see Part 13 – CORGI*direct* Publications.

*   any control interlocking device is set to its provisional operating level, considered safe for commissioning (e.g. regulator, process controls and interlocks)

*   with auxiliary energy supplies available, all power equipment and interlocks are checked for operation, e.g. where automatic control units are employed, the controller is checked at each stage of its light-up sequence for correct operation and response of the control unit

*   check that the system of ignition is adequate, e.g. strength and location of any spark generation is correct to ensure smooth and reliable lighting of start gas flame

*   flame safeguard systems correctly go to shutdown condition in the absence of a flame, e.g. lockout within the specified time period. If a simulated flame is present the system detects the presence of the flame correctly.

Activation – fuel run

Once all checks have been made with the fuel isolated to the satisfaction of the commissioning operative, the gas may be turned on and the following further checks initiated:

*   the appliance is correctly purged of air (pipework connector, appliance gas train and combustion chamber)

*   allow the burner system to proceed through to start gas ignition or pilot stage, ensure that start gas flame or pilot is established and set at the correct pressure before proceeding to main gas

*   confirm by suitable means of leak detection that all start gas or pilot gas pipework is gas light

*   ensure that loss of start gas/pilot flame results in the correct shutdown of the control system. Check thermo-electric devices for the correct 'drop-out' time as described in manufacturers information

*   ensure that, with the burner shutdown, the safety shut off valve(s) remain gas tight

*   with main gas available and controls restricted to a nominal main flame ignition rate, allow the burner to proceed to main gas ignition. Ensure that the main gas flame is established and set at the correct pressure

*   confirm by suitable means of leak detection that all main gas burner pipework is gas light

*   ensure that loss of gas flame results in the correct shutdown of the control system

*   ensure that, with the burner shutdown, the safety shut off valve(s) remain gas tight.

Operational checks

Once all checks have been made to ensure safe and reliable ignition to main gas burner stage, to the satisfaction of the commissioning operative, the following further operational checks can be initiated:

*   the heating plant is allowed to run up to normal operational temperatures to ensure the unit remains satisfactory

*   operational controls can now be checked for satisfactory operation, e.g. room thermostats

*   combustion gases are effectively dispersed, ventilation is adequate

- combustion checks are carried out, which dependant on the manufacturer's instructions may require the use of a portable electronic combustion gas analyser (see Note)

- undertake spillage tests, where appropriate

- ensure that the requirements of the GSIUR, Regulation 26(9) are met, i.e. the effectiveness of any flue; the supply of combustion air; its operating pressure or heat input or, where necessary, both; its operation so as to ensure its safe functioning.

Note 1: When undertaking combustion analysis in accordance with the manufacturer's instructions, the correct instrumentation will need to be used ensuring that it is within calibration by having a valid calibration certificate; it is suitable for the application; the operative is familiar with the instrumentation and how to interpret the readings and therefore, ensuring correct levels of accuracy are achieved.

Note 2: For further detail on the correct procedures for combustion analysis and appliance efficiency testing, see 'Essential Gas Safety – Non-Domestic' (Order Ref: ND1) and 'Combustion performance testing - non-domestic' pocket guide (Order Ref: CPA2), see Part 13 – CORGI*direct* Publications.

The industry standard referred to generally for using ECGAs in non-domestic premises is BS 7967-5: 2010 'Carbon monoxide in dwellings and other premises and the combustion performance of gas-fired appliances. Guide for using electronic portable combustion gas analysers in non-domestic premises for the measurement of carbon monoxide and carbon dioxide levels and the determination of combustion performance'.

Additional considerations for direct-fired heaters

The general commissioning tasks, including servicing discussed in **Part 10 – Servicing, maintenance and fault finding** will be as described for indirect-fired heating systems. However, there are some additional requirements for flueless appliance systems described here.

It will be necessary to ensure that POC do not build-up beyond accepted limits within the heated environment. If the direct-fired system has been correctly designed and is part of a make-up air system, the warm air distributed from the heater is replacing contaminated air being extracted.

BS 6896, Annex A 'Testing for carbon dioxide in the environment', describes the maximum CO and CO_2 concentrations at any point where persons can normally be expected to work. This includes low levels, mezzanine levels and at high levels.

At these points CO_2 concentration should not exceed 2800 ppm and CO limited to 10ppm.

Checks should be undertaken when the building is operating under full heat input conditions, usually after at least the first hour of operation.

The HSE provide further guidance on exposure limits for various substances that can be encountered in the workplace via EH40/2005 'Workplace exposure limits'. For details of these limits, refer to **Principles of Warm Air Heating – Free discharge systems – direct-fired** previously covered in this Part.

To undertake these tests, the correct instrumentation will need to be used to ensure it is suitable for the application and correct levels of accuracy are achieved.

For example, it will not usually be appropriate to use typical portable electronic combustion gas analysers (ECGAs) to sample CO_2 in the atmosphere - those analysers complying with BS EN 50379-4: 'Specification for portable electrical apparatus designed to measure combustion flue gas parameters of heating appliances. Performance requirements for apparatus used in non-statutory servicing of gas-fired heating appliances' – as they do not measure CO_2 directly, but merely calculate a CO_2 level from the directly related excess oxygen measured in the sample.

Where measuring CO_2 directly, the ECGA should comply with BS 8494: 2007 'Electronic portable and transportable apparatus designed to detect and measure carbon dioxide in indoor ambient air. Requirements and test methods'.

Completion

Once all aforementioned checks have been made to the complete satisfaction of the commissioning operative the following will be required:

- all users are instructed in the correct operation of the heater and/or system and its user controls, light-up and shutdown sequence

- manufacturer's 'user' instructions are left for the responsible person. The 'responsible person' is defined by the GSIUR as the 'owner or occupier'. However, on many large commercial or industrial sites this may not be relevant, it may be more appropriate to identify the works engineer or manager for this purpose

- the responsible person should also be advised concerning matters such as actions to be taken in the event of fault or emergency conditions and advice concerning regular maintenance of the units

- a suitable report will need to be completed and left with the responsible person. This report needs to detail the final setting parameters of the heater/system, including:

  - customer/site details

  - plant/heater details, e.g. make, model and serial number

  - fuel supply details, e.g. type and supply pressure

  - appliance/system operating set levels

  - combustion and emission data

  - electrical data, e.g. nominal supply, overload settings, fuse ratings.

Note: CORGI*direct* **provides suitable forms for the correct documentation of checks/test carried out, which gas operatives' can provide to their customers. The forms, 'Plant Commissioning/Servicing Record' (Order Ref: CP15), 'Gas Testing and Purging' (Order Ref: CP16) and 'Gas Installation Safety Report' (Order Ref: CP17) are all detailed in Part 13 – CORGI***direct* **Publications.**

Handover

Upon completion of all the aforementioned commissioning actions, the commissioning report, together with any manufacturer's installation and maintenance instructions are handed over to the designated responsible person for the site/premises.

It should then be made clear to all concerned that the commissioning process has been completed and responsibility for the plant is passed on to the appropriate personnel.

Combined heat and power – 9

9 – Combined heat and power

Introduction

Electrical power is conventionally produced by a generator driven by a heat engine; this heat engine could be a steam turbine in a power station or an internal combustion engine for smaller and mobile power generation.

In conventional power generation approximately 65% of the primary energy required in the power generation process will be rejected, mostly in the form of water at a nominal temperature of 30°C to 35°C, but also in the form of hot exhaust gases from the engine.

By using heat exchangers in the system it is possible to recover most of this wasted energy. Approximately 33% of the heat can be recovered from the engine jacket at about 120°C, which can be used to generate hot water at 70°C to 85°C; approximately 22% from the exhaust gases at 650°C can be recovered through a gas-to-water heat exchanger and provide central heating.

In this way, the energy used in the power generation process can be fully utilised to provide combined heat and power (CHP) for a particular building or factory.

The use of NG as the source fuel for the power generating system has become increasingly common over recent years for all the advantages that running on a clean efficient sulphur free fuel has to offer.

Fundamental to this process however, is the correct design based on the heat to power ratio. To ensure that the waste heat generated from the power generation process can be fully utilised there will need to be ample requirement at any given site for heating and process hot water. If not, the building management systems for the site will be forced to reject the hot water, usually by discharging to waste sewage or the national river systems.

In either case, approval will need to be sought from the appropriate water and/or river authority at significant financial cost, as it will probably require cooling through a purpose provided cooling tower or fan cooled radiator.

CHP systems will therefore usually be designed to satisfy the constant base load for power, heating and hot water at any given site with conventional systems used to 'top up' and match any load requirement in excess of the base demand.

Alternatively, where the demand for heating and hot water is greater than the need for power at any given site, it may be possible subject to contract, appropriate metering facilities and prior agreement with the power transmission companies, to supply excess electrical power back into the national power grid system (referred to in this manual as 'grid').

General

Changing economics of energy management, together with legislative and regulatory initiatives, particularly over most recent years, have resulted in increased awareness of on-site power generation technology and an ever increasing demand for CHP systems.

CHP systems are generally divided into 3 basic categories, namely:

- micro (also termed 'micro-cogeneration')

- small-scale

- large-scale.

Much of the development that has taken place over recent years is at the small-scale and micro-CHP end of the market.

In the gas utilisation market, small-scale CHP is usually based on gas-fired internal combustion engines producing electrical outputs of up to 1MW.

Figure 9.1 Typical variable load fluctuation

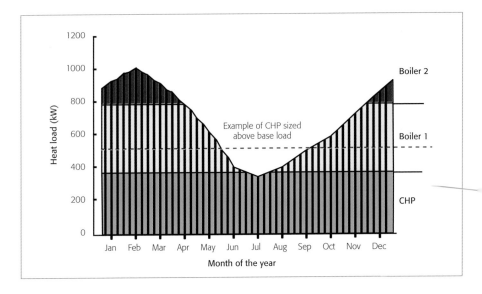

Whilst Micro-CHP represents the smallest of that category for use in domestic or small commercial applications with electrical outputs from as low as 5kW.

The use of small scale CHP systems have proved to be an energy efficient, cost effective development for a wide range of buildings such as leisure centres; swimming pools; hospitals; universities etc.

CHP should not be considered in isolation for satisfying the energy demands at any given site or building, but as an integrated part of the total energy system.

The constant power, heating and hot water base demand being fed by the engine, whilst conventional heat and power is provided to match the variable demands of the days and seasons.

Figure 9.1 illustrates diagrammatically how energy demands can be met using a CHP unit in conjunction with traditional heating systems (e.g. boilers) to match the demand profile.

Most small-scale CHP units provide LTHW and can therefore be directly interconnected with standard boiler plant. Figure 9.2 illustrates how the CHP unit may be systematically interlinked, either in series, or in parallel with modular boilers to match the load profile illustrated in Figure 9.1.

CHP units will often be connected in series with modular boilers on existing heating systems, whereby one of the boilers can be replaced by the CHP unit. CHP units connected in parallel will be more appropriate if the CHP unit is supplying a significant proportion of the heating load.

Figure 9.2 Typical CHP plant connected to conventional heating system

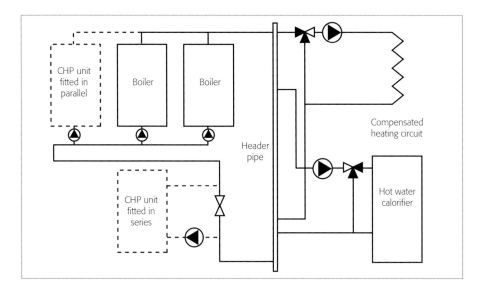

Principle of CHP

CHP systems can run on NG, LPG, gas-oil or bio-gas and the range of units available for buildings is generally as follows:

- micro – up to 5kW electrical power

- small-scale – below 2MW electrical power with various engine systems, e.g.

    - spark ignition engines

    - micro-turbines producing between 30 and 100 kW of electrical power

    - small-scale turbines producing approximately 500kW of electrical power

- large-scale – above 2MW electrical power

    - large reciprocating engines

    - large turbines.

Most small-scale CHP units come as a complete packaged unit complete within an acoustic housing to minimise noise transmission.

The system will comprise of a NG fuelled spark ignition gas engine, which is used to drive an electrical generator. Heat is recovered from the exhaust and the water cooling systems (see Figure 9.3).

Components of CHP

Figure 9.4 shows the typical components of a CHP unit.

Reciprocating engine

Most packaged small-scale CHP units utilise a spark ignition gas reciprocating engine as the prime mover. These units typically produce electrical power up to approximately 800kW with a heat to power ratio of 1.5:1. So, typically a unit producing 100kW electrical power will also produce 150kW heat output.

Figure 9.3 Typical packaged CHP unit

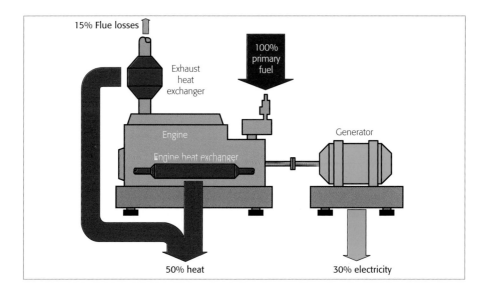

Figure 9.4 Typical components of a CHP unit

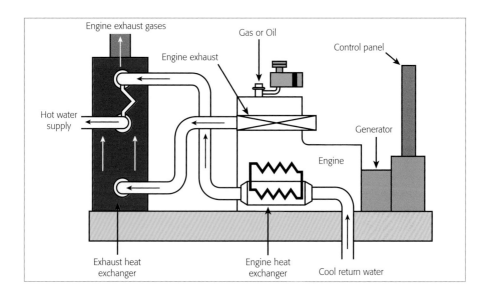

Figure 9.5 Example of turbine generator system

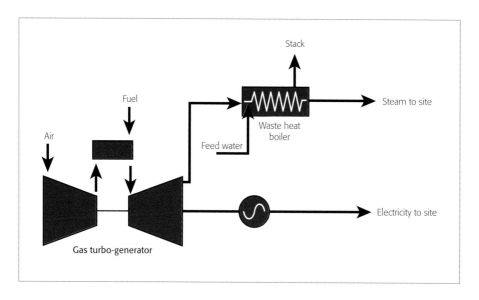

These engines generally have higher shaft efficiencies than turbines and can usually perform better on part loads. Most of these units are capable of modulating down to approximately half of full load and still maintain good operating efficiency.

The use of NG as the source energy has increased over recent years as a result of various factors, such as government directives to industry and commerce for improved energy efficiencies and improved unit availability resulting from market forces. But underpinning all of this is the highly desirable requirement that the gas burns cleanly with minimal carbon deposits and no dilution of lubricant oil.

Gas turbines

Gas turbines are typically used on larger power output units, e.g. above 1MW electrical power, for industrial power generation.

Turbines utilise combustion gases fired at high pressure to rotate a series of spiral blades mounted on a rotating shaft, the shaft being attached to the generator. The same turbine or a separate power turbine is used to drive an air compressor, which delivers the air required for combustion of the gas (see Figure 9.5).

Combustion gases are delivered to the turbine at temperatures between 900°C and 1200°C, which will then exhaust at approximately 500°C.

This higher output exhaust is ideally suited for high-grade heat supply for industrial processes and heat to power ratio can be between 1.5:1 and 3:1 dependent on the particular turbine.

In addition, because combustion always takes place with excess air, the hot excess air exhausted can be recycled to provide a means of recuperated air supporting combustion in other processes. Where this can be effectively utilised the heat to power ratio can increase up to 5:1.

Another advantage of the gas turbine system is that the constant high speed will enable a closer frequency control of electrical output.

Recent development in gas turbines has produced the micro-turbine technology, which means that packaged turbine units are now available for applications as low as 30kW electrical power.

Overall efficiencies are comparable to reciprocating engines, as long as all the process advantages of using turbine technology can be fully utilised, i.e. the ability to utilise high temperature exhaust and recuperation.

Generator

'Synchronous' generators are usually used for stand alone units as the output voltage and frequency are determined by the generator control unit; they can modulate output power over a wide range, do not require power factor correction and are designed to operate in isolation from the grid. They can therefore be used as standby generators.

'Asynchronous' generators are usually used on Micro-CHP units; they are mains excited and cannot therefore operate independently of the grid.

They operate in parallel with other generators, such as the grid and so the grid determines the output voltage and frequency. Also if the grid supply fails the generator output stops.

Mains excited asynchronous generators are identical to an induction motor and can therefore be used to start the engine. When mains excited, the motor can rotate the engine until the engine is up to speed. Once up to speed, the engine becomes the driving force and the role of the motor changes to that of generator.

Heat recovery

Heat is mainly recovered from the waste processes of power generation using a number of heat exchangers in the engine cooling water and the exhaust gases. Additional heat can sometimes be recovered from the exhaust gases by using a condensing economiser and/or from heat exchangers in the lubricating oil system and the generator and the unit housing. However, these additional heat exchangers will only be fitted where there is economic justification.

Types of heat exchangers used will vary dependent on the waste heat source, but will be:

• water-to-water for heat from engine jacket and cooling system

• gas-to-water for heat from exhaust and manifold gases

• air-to-air for turbine recuperation systems.

Whatever heat exchangers are fitted they represent replaceable items that will on occasion require appropriate maintenance and their life expectancy will probably be less than the engine; so they need to be relatively easy to access and repair/replace.

The maximum operating water temperatures and pressures for recovering waste heat are generally similar to those used for other LTHW systems, e.g. nominally 70°C – 85°C and 3 to 7 bar.

Where condensing heat exchangers are fitted in the exhaust gases to drop them below the dew point at approximately 55°C, they are likely to be constructed of stainless steel or perhaps aluminium.

Controls

Firstly, the control requirements for the CHP unit will be similar to a conventional boiler system, i.e. configured to preheat the return water to the boiler or be operated as a lead boiler in parallel on a modular system.

Control in relation to electrical load will be part of the packaged system and will incorporate continuous monitoring as part of the control system.

The data output from the control system can be used to form the basis for planned maintenance such that work can be carried out prior to any breakdown.

The functions of the control unit are varied and include:

- controlling start-up and shut-down sequence
- automatically control power output to meet variable load conditions
- integrate with other heat and power systems.

The safety controls fitted to the engine need to fulfil the following requirements:

- automatically shut down safely in the event of particular system faults such as –
  - engine over-speed
  - low engine oil pressure
  - high and low gas supply pressure
  - gas leakage (in the event of minor gas escapes on unmanned plant)
  - high water temperature
  - electrical over voltage
  - electrical over frequency
  - electrical reverse power.

Installation requirements

The following information provided in this Part offers general guidance notes on the types of CHP units previously described. Specific information relating to the installation of any particular CHP system will be described in the appropriate manufacturer's instructions.

Particular installation normative reference documents include:

- IGEM/UP/2 (Edition 2) 'Installation pipework on industrial and commercial premises'
- IGE/UP/3 (Edition 2) 'Gas fuelled spark ignition and dual fuel engines'
- CIBSE Applications Manual: AM12 1999 Small-scale combined heat and power for buildings
- EH40/2005 Workplace Exposure Limits Containing the list of workplace exposure limits for use with the Control of Substances Hazardous to Health Regulations (as amended) – available freely as a pdf download from www.hse.gov.uk/pubns/books/eh40.htm or as a hard copy from HSE Books

In order to assist gas operatives, and others concerned with the installation of gas appliances in to non-domestic premises, the Communities and Local Government (CLG) produce a free guide, entitled 'Non-Domestic Building Services Compliance Guide', which provides the reader guidance on types of heating systems and relative efficiencies required when installing heating in new or existing premises.

Note: The guide can be obtained by visiting http://www.planningportal.gov.uk/buildingre gulations/approveddocuments/partl/bcassoc iateddocuments9/compliance

For further information regarding the Building Regulations see **Part 2 – Gas and associated legislation – Building Regulations (England and Wales)** in this manual.

When designing any CHP system there are many areas to consider; collaboration will be essential between those involved such as the gas operative, the CHP manufacturer and system designer, the occupier and the building owner. There may also be the need to inform local authorities for appropriate planning consent (e.g. for listed buildings), fire authorities, building insurers etc.

Additionally, the installation will need to take in to account the requirements of Building Regulations applicable to the geographical area concerned (see **Part 2 – Gas and associated legislation** for further guidance).

Matters to be considered include:

- designers requirements

- manufacturer's installation requirements

- building construction (fabric heat losses, suitability and proximity of materials etc.)

- flueing requirements

- ventilation requirements

- gas availability
  (type, site capacity and pressure)

- electricity requirements

- site access and liaison with associated trades, etc.

Note: The above list should not be regarded as exhaustive.

Competence

Work carried out relative to the installation of CHP units needs to be undertaken by operatives who are experienced in such work and who hold the relevant certificates of competence obtained through the ACS and/or aligned N/SVQ's.

Note: For more information regarding commercial ACS categories refer to Part 3 – 'Competence' and Essential Gas Safety – Non-Domestic (Order Ref: ND1), see Part 13 – CORGI*direct* Publications.

General location requirements

Manufacturer's instructions will provide clearance requirements for adequate maintenance access. All items that are liable to wear or otherwise require regular attention will need adequate access provision to allow withdrawal/removal without causing damage to other components on the unit or system.

If the engine is not contained within its own housing, due consideration will need to be given to the effect of the heat generated on surrounding building fabric so as not to cause degradation of that fabric; where necessary, suitable shielding will need to be provided accordingly.

Manufacturers will also provide advice on floor loadings to ensure the correct foundations are built for the unit and associated system.

If the unit is to be located within a building with atmospheres containing vapours, gases or airborne dust, then it will need to be located within its own housing, adjacent to external walls to allow fresh air to be ducted to the engine for combustion.

Any permanently installed CHP unit will need to be securely located to prevent excessive displacement. However, when necessary, anti-vibration mounts will need to be fitted; this is to minimise damage caused by vibration.

Condensing units will involve additional considerations particularly regarding the condensate removal. The condensate disposal pipe will need to run indoors to avoid the possibility of freezing in winter. The location of the unit can sometimes be dictated by the nearest access to soak away drainage facilities.

Also, condensing exhaust systems can have a tendency to form a plume of water vapour from the flue terminal; consideration will therefore be required as to the effect of this pluming on surrounding building structure, materials and neighbouring buildings.

When designing or considering an existing area for the engine room, explosion relief panels may need to be fitted. In general, explosion relief will not usually be required, but IGE/UP/3 (Edition 2) will need to be consulted for specific advice.

Restricted locations

Individual CHP units and associated systems will need to be installed in locations that are deemed suitable. In industry particularly, there will be a wide variety of applications and processes, therefore, due consideration needs to be given to the likely environmental conditions, for example:

- corrosive or salt-laden atmospheres will obviously have an affect on metallic components, particularly burners and controls, e.g. metal plating/treatment workshops

- dusts and vapours given off from plastic forming processes, cleaning or curing applications when passed into the atmosphere will be drawn into the combustion chamber and may subsequently produce harmful gases

- chlorine laden atmospheres (e.g. at swimming pools/baths) – when atmosphere is laden with chlorine in the vicinity of a permanent heat source, the chlorine will break down to base component chemicals, including acids which will attack the metallic components of the heater and corrode very rapidly

- LPG fuelled engines cannot be located in a totally enclosed room below ground e.g. cellar or basement

- the engine needs to be located in a separate enclosure to any compressor for flammable gases.

Where the engine is the driving force for a flammable gas compressor the two machines will need to be located in separate enclosures.

In this situation the drive shaft between engine and compressor will either need to have a guaranteed sealing mechanism or the enclosures will need to be located with a physical externally vented air gap between them

This may also apply to a refrigerant compressor if driven by an engine, due to the possibility of the refrigerant being corrosive if drawn into the engine air supply.

If there is any doubt as to the suitability of any location for a given application, always seek advice from the manufacturer(s). Most manufacturers will prefer to give, often free advice, in order to ensure the unit is appropriate for the application, rather than risk the possibility of inheriting bad publicity resulting from unsuitable/poor installations.

Air supply/ventilation

To achieve complete combustion it is essential to ensure that the appliance is supplied with an adequate supply of clean fresh air (ventilation).

This will necessarily take into account the type of chimney/flue pipe fitted to the appliance and any other appliances or other fuel-fired plant in the same area, along with any cooling requirements for the appliance and the occupants of the workspace in which the appliance is sited.

Any ventilation provision(s) needs to be in accordance with the manufacturer's specification. In addition to manufacturer's instructions, the industry standard referred to for ventilation installations, relative to gas fuelled spark ignition engines systems is IGE/UP/3 (Edition 2).

The amount of air will need to be determined for each individual application and will comprise the sum of air required for combustion, cooling and permanent ventilation for the engine when it is running and when it is shut down.

Combustion air will be required direct to the air filter inlet in accordance with manufacturers' instructions.

Cooling air will need to be provided to cool the engine when running, based on a number of factors:

- engine rating

- heat emitted from other adjacent plant including associated CHP equipment

- method of cooling, e.g. fan cooled radiator system or water-to-water heat exchanger

- method of heat recovery from the exhaust system

- quality of exhaust insulation

- whether the engine is enclosed and if so, size and form of the construction

- heat emissions from other machinery driven by the engine.

In any event the cooling air requirement will need to comply with the manufacturers' stated maximum ambient temperature. High ambient temperatures, e.g. in the summer, may affect engine output if above the designed maximum operating temperature.

Ventilation air needs to be provided to ensure that when the engine is shut down there is still a sufficient flow of air to prevent any possible build-up of combustion gases in the event of minor gas leakage in the engine room.

For lighter than air gas (NG) engines this will require ventilation at high level, for heavier than air gas (LPG) engines ventilation will be required at both high and low level.

If the provision of an air supply requires the operation of a separate mechanical fan system, the fan system will need to be interlocked to prove airflow prior to engine start-up and during engine running mode.

However, any such interlock will need to be configured in such a way as to ensure the fans continue to run for a period of time after the engine has shut down for engine cooling purposes.

In certain cases, the supply of ventilation air can be dependent upon power from the engine to operate the fan system. Where this situation exists, a risk assessment will need to be carried out to establish whether or not a build up of gas if/when the engine is idle could lead to a hazard. If so, a system of gas detection will need to be fitted to prevent engine start up if gas leakage is indicated.

Note: for further guidance on risk assessments see Essential Gas Safety – Non-Domestic, (Order Ref: ND1), see Part 13 – CORGI*direct* Publications.

IGE/UP/3 (Edition 2) gives further guidance on the type and location of gas detection instrumentation.

Exhaust system

The manufacturer's exhaust adaptor piece should always be used to ensure correct connection to the exhaust spigot without reducing the cross sectional area (csa) and that a complete seal is obtained. Consider also, that provision will need to be made so that the exhaust can be disconnected from the CHP unit to allow inspection and servicing.

Exhaust pipework will need to be constructed of appropriate material with welded or flanged jointing systems capable of withstanding the temperatures and pressures expected from the engine. The pipe route will need to be as short as practical whilst ensuring the complete and safe removal of the POC to atmosphere.

Where heat dissipation from the exhaust system is undesirable or represents a hazard to personnel in the area, the system will need to be lagged with appropriate insulation material.

Given that the engine will be prone to some vibration, manufacturers' advice will need to be sought as to the most appropriate exhaust connection to avoid transmission of the vibration to the exhaust system.

The exhaust system will need to be of sufficient strength to withstand any sudden internal pressure rise as may occur from back firing of unburnt gas/air mixtures; usually mild steel is adequate. However, weaker flexible pipe sections may represent a particular hazard. In which case a system of explosion relief such as a wire cage, to retain fragments in the event of fracture will need to be installed.

The route and design of the exhaust system will need to be such that pockets of unburned gas cannot accumulate. In the case of NG the exhaust system will need to rise continuously to allow natural venting of the gas. Where this is not possible, a system of forced or induced purging will need to be considered.

The exhaust discharge terminal will need to be positioned so as to ensure that no POC are allowed to re-enter the building or re-circulate into the combustion air intake, but also be located where they cannot be easily blocked by water ingress or other foreign objects.

Ideally the CHP unit will require a dedicated exhaust system, however, where it is unavoidable to use a common flue system with other engines or plant, the engine exhaust will need to be the highest connection into the flue. Particular attention will need to be paid to the flueing requirements of other gas-fired appliances, e.g. modular boilers, to ensure that the engine exhausts do not cause backflow of the POC in the flue to the other shared appliances.

Where condensation in the flue is unavoidable, appropriate provision will need to be made for the condensate to flow freely to a disposal point. Also, particular attention will need to be paid in respect of the materials used for the exhaust, especially if the engine is dual fuelled; gas oil will have a level of sulphur in the condensate, higher than that of NG and therefore be more corrosive.

Where non-metallic exhaust pipe is used for condensing flue systems, the following are some of the essential requirements:

- the pipe system is gas tight and capable of withstanding up to 1.5 times the normal working pressure of the engine exhaust

- the jointing system of the pipe will need to be either flanged or solvent welded. Push connectors and compression joints should not be used

- the route of the pipe will need to be such as to avoid risk of mechanical damage and heat transfer from other plant that would increase the surface temperature of the pipe system above 50°C

- if flue gas temperatures above 50°C are expected then the pipe system will need adequate support to prevent sagging

- an additional engine shut down control may be required if the POC entering the non-metallic system is greater than 60°C.

Additionally, requirements for condensate disposal will be as follows:

- means of drainage will need to meet approval of local water authority

- materials will need to be suitable for the condensate, i.e. not copper or copper alloy

- diameter of the drain will need to be at least 20mm diameter

- means of condensate collection will need to be located at all low points in the system and be water tight to prevent leakage of exhaust gases.

Gas installation

The GSIUR require gas fitting operatives to ensure that gas installation pipework and fittings are installed safely with due regard to the location of other services, e.g. other pipes; pipe supports; drains; sewers; cables; conduits and electrical control equipment.

Gas operatives will also need to be aware of the limitations of the building structure when installing gas equipment and pipework.

Note: For information relating to the general requirements for gas pipework in non-domestic establishments see Essential Gas Safety – Non-Domestic (Order Ref: ND1), see Part 13 – CORGIdirect Publications.

The industry standards referred to for non-domestic gas pipework installations feeding CHP systems are IGE/UP/2 (Edition 2) and IGE/UP/3 (Edition 2).

Also, in recognisance of the fact that engine gas supply pressures may be required in excess of 500mbar, the Pressure Systems Safety Regulations 2000 (see **Part 2 – Gas and associated legislation**) will also need to be complied with.

It is particularly important for the gas pipework feeding any CHP unit to be completely clean internally and free of loose debris, e.g. welding slag. Any cleaning process will need to be applied prior to commissioning and also after any subsequent pipework repairs or alterations are completed.

In any event the need for a suitable filtration system in the gas supply will be essential, taking into consideration also the potential for erosion that could take place within the pipe and filter where high velocities will be encountered, typically having a mesh of 10 – 50µm.

An emergency isolation valve of a manual operation type will need to be fitted and correctly labelled at a point outside the engine room or enclosure. In addition, where considered necessary, this manual valve may have a remote actuation facility.

In any event, a quick acting manual valve will need to be fitted immediately upstream of each engine gas control train. This valve could also act as the emergency isolation valve if located externally to the engine enclosure.

Gas connections

It is reasonable to expect that the engines will be subjected to a certain amount of vibration. Guidance should be sought from the engine manufacturer as to the need for gas flexible connections. Where a flexible connection is required it will need to be as short as possible whilst allowing sufficient engine movement.

Where manufacturers stipulate that the final gas connection to the CHP unit is by means of a flexible connection, that connection needs to be constructed out of suitable material.

Stainless steel for example, complying with BS 6501-1: 2004 'Metal hose assemblies. Guidance on the construction and use of corrugated hose assemblies' and BS EN ISO 10380: 2003 'Pipework. Corrugated metal hoses and hose assemblies'.

For added protection, it is advisable that the flexible connection incorporates a surface cover. This may be a plastic sleeve, or if there is risk of physical damage, over-braided stainless steel.

Note: The decision as to the appropriate sleeving material to use will need to be made, bearing in mind the environment in which the CHP unit and therefore, its gas connection will be exposed to. For example, BS 6501 strongly recommends that plastic coatings containing either sulphur or chlorine should not be used.

Any gas flexible connection needs to be installed such that there are no sharp 90° bends and ensuring the connection is not subjected to stress or torsion.

A flange joint will be required at one end of the flexible connection to facilitate disconnection of the gas pipework for maintenance of the CHP unit. The gas operative should ensure that the flexible connection does not twist when the pipe is reconnected and tightened.

The flexible connection will need to be installed immediately downstream of at least one of the safety shut off valves (SSOVs).

In certain cases, particularly where the supply pressure is 75mbar or more, due consideration will need to be given for automatic means of preventing the CHP unit and associated pipework from being subjected to any pressure higher than the pressure it is designed to operate at, e.g. in the event of regulator failure. This 'protection' would usually be by means of a regulator with integral overpressure slam-shut valve incorporated.

Additional components that may be encountered fitted in the gas train to the engine include low gas pressure cut off switch interlocked to the SSOV system, non-return valve and adequate test/purge points.

Electrical connections

Non-domestic properties normally utilise three-phase electrical systems due to the high current demands of the electrical equipment installed and used.

When gas operatives are required to work on gas appliance/equipment, which utilise either three-phase or single-phase electricity, it is important that those gas operatives are competent and have the relevant knowledge and experience to work safely on those systems.

They need to be able to carry out a thorough risk assessment of any potential hazards that may be involved. For further information on risk assessments see Essential Gas Safety – Non-Domestic – Part 3 – Risk assessments.

Important: The Electricity at Work Regulations 1989 (EWR) apply to any operative carrying out any type of electrical work. It is a requirement of the EWR that the person is competent to undertake the work.

Where any doubt exists with regards to competency or lack of knowledge/experience to work on those electrical systems, or where a risk assessment identifies that the hazard would be unacceptable, no work should be undertaken.

Working on three-phase equipment, supplies and four wire supplies

In the UK the 400V, 4-wire three-phase and neutral (TPN) method is widely used for the distribution of supplies within commercial and small industrial installations.

For example, on large wet central heating systems the circulating pump may operate at 400V from a three-phase supply while other components such as thermostats and time switches may be 230V single-phase or sometimes even less, for example 24V, which is classed as Extra Low Voltage (ELV).

On commercial warm air heaters, the fans (combustion and/or circulating) may be three-phase, while other components may be single-phase.

It must be remembered that each of these phases is carrying 230V. A balanced three-phase circuit does not require a neutral for it to be able to operate. This is because there is always a return path for the current through one of the other two phases.

There are a number of signs to look for when trying to identify the voltage on non-domestic gas installations, two most basic signs are:

1. Appliance data plate – one of the simplest methods of determining the voltage to a non-domestic gas appliance is by checking the information on the appliance data plate, assuming of course a plate is present.

    The information displayed will typically consist of:

    -   supply voltage – 400V

    -   supply frequency – 50Hz

    -   no of phases – 3

    -   no of wires – 4

    -   power consumption – 4.0kW

    -   fuse rating – 400V, 25A HBC

    -   fuse rating – 230V, 5A

2. Installation wiring – all installation wiring falls within the scope of BS 7671: 2008 (Including Amendment No.1) 2011: 'Requirements for electrical installations. IET Wiring Regulations Seventeenth Edition' and therefore requires completion of electrical certification when installed.

    The existence of a three-phase supply can often be confirmed by the wiring colour code, this will need to be tested to ensure it is correctly sized etc. before undertaking any electrical installation work.

Remember, all electrical installations should be in accordance with BS 7671 and the supplied manufacturer's wiring guides, especially for the correct connection of supplies.

Of equal importance is the correct routing of cables, including suitable supports where required and the correct securing of cables to individual pieces of equipment.

Each appliance and its associated controls needs to have its own means of electrical isolation, which should be situated in close proximity to the equipment is serves and have suitable means for the application of securing devices, i.e. hasps or padlocks for ensuring that the appliance cannot be re-energised whilst being worked on by an operative (this should be highlighted as part of a risk assessment).

Commissioning

Manufacturer's information will provide specific information for commissioning a particular CHP unit and its associated equipment.

In addition, IGE/UP/3 (Edition 2) and IGE/UP/4 (Edition 2) 2nd Impression 'Commissioning of gas-fired plant on industrial and commercial premises' offer further generic guidance for the commissioning process.

Correct commissioning of an appliance in accordance with the manufacturer's instructions is as critical to a safe and efficient installation, as its initial installation.

Therefore, where the installation can not be commissioned immediately after installation, it should be isolated from the gas supply and suitable notification attached (this fact also needs to be documented on any appropriate paperwork) to advise the reader that the installation is un-commissioned.

The following is provided for information purposes only and should not be regarded as substitute for the source documents.

Planning

Before travelling to site, a certain amount of preparation will be necessary, including:

- ensure all relevant manufacturer's information is available

- all relevant drawings and system plans are available

- all gas tightness test/purge certificates have been completed by the upstream pipework installer

- all electrical test certificates have been completed by the electrical installer

- risk analysis has been completed

- COSHH statements prepared, where relevant

- any 'hot work' or 'permit to work' has been granted (see Essential Gas Safety – Non-Domestic – Part 3 – 'Risk assessments' (Order Ref: ND1), see **Part 13 – CORGI***direct* **Publications**)

- access is available

- all necessary tools are available and suitable for the purpose and are of sound quality. Any electronic instruments need to be correctly calibrated and certificated accordingly.

- a written plan of required work is completed (usually provided in manufacturer's instructions).

Inspection

Once on site, a complete inspection of the installation will need to be carried out to ensure that:

- all gas and electrical supplies are suitably sized, correctly located and supported and that the installation has been installed in accordance with the manufacturer's instructions and relevant standards

- assembly is complete and all components are fit for the intended purpose

- the CHP unit has been correctly supported on a suitable base; it is level as required by manufacturer's instructions and is stable with adequate clearance for maintenance and from combustible materials

- gas and electrical supplies have been isolated.

Activation – dry run

Subject to satisfactory inspection, the next phase of the commissioning process is to carry out preliminary physical checks with electricity and fuel isolated:

- a suitable gas tightness test is undertaken on the appliance gas connection and its isolation valve(s)

Note: For detailed guidance on the correct procedures for testing gas pipework in non-domestic establishments, see Essential Gas Safety – Non-Domestic (Order Ref: ND1), see Part 13 – CORGI*direct* **Publications).**

- SSOVs are checked for 'let-by' on both sides of the valve. Each valve will need to maintain a tight shut off against all reverse differential pressure and will need to be tested to the greater of the following:

  - twice maximum air pressure at the point of fuel injection to the manifold or cylinder, or

  - 150mbar for SSOVs not exceeding 50mm bore and 100mbar for SSOVs in excess of 50mm bore

- any engine unit of rated input in excess of 1.2MW will be fitted with automatic SSOV proving system; check system for correct operation

- any non-return valve is checked for correct operation

- all control interlocks are checked for correct operation, e.g. high and low gas pressure switches

- all control interlocking devices are set to a provisional operating level, considered safe for commissioning, e.g. regulator, process controls and interlocks

- with auxiliary energy supplies available, all power equipment and interlocks are checked for operation, e.g. where automatic control units are employed, the controller is checked at each stage of its sequence for correct operation and response of the control unit

- check that the system of ignition is adequate, e.g. strength of spark generation is correct.

Activation – fuel run

Once all checks have been made with the fuel isolated to the satisfaction of the commissioning operative, the gas may be turned on and the following further checks initiated:

- the pipework to the engine is correctly purged of air

- allow the engine to run to ignition and is fully commissioned in accordance with manufacturer's instructions.

Operational checks

Once all checks have been made to ensure safe and reliable ignition, to the satisfaction of the commissioning operative, the following further operational checks can be initiated:

- the CHP plant is allowed to run up to normal operational temperatures to ensure the unit remains satisfactory

- the engine is set to the correct run rate. The commissioning engineer will need to ensure that in bringing the engine up to its operating running rate the manufacturers required 'ramp up' rate is not exceeded, i.e. the rate at which the engine is raised to its required operating running speed

- with the engine running, check all gas pipe downstream of the shut off system and engine mounted fuel lines for tightness using leak detection fluid or suitable gas detection instruments

- the air/gas mixture is adjusted to achieve optimum performance

- combustion gases are effectively dispersed, ventilation is adequate

- combustion checks are carried out, which dependant on the manufacturers' instructions may require the use of a portable electronic combustion gas analyser (ECGA), see Note

- operational controls can now be checked for satisfactory operation, e.g. control thermostats

- undertake flue flow and spillage tests, where appropriate

- ensure that the requirements of the GSIUR, Regulation 26(9) are met, i.e. the effectiveness of any flue; the supply of combustion air; its operating pressure or heat input or, where necessary, both; its operation so as to ensure its safe functioning.

Note 1: When undertaking combustion analysis in accordance with the manufacturer's instructions, the correct instrumentation will need to be used ensuring that it is within calibration by having a valid calibration certificate; it is suitable for the application; the operative is familiar with the instrumentation and how to interpret the readings and therefore, ensuring correct levels of accuracy are achieved.

Note 2: For further detail on the correct procedures for combustion analysis and appliance efficiency testing, see 'Essential Gas Safety – Non-Domestic' (Order Ref: ND1) and 'Combustion performance testing - Non-domestic' pocket guide (Order Ref: CPA2), see Part 13 – CORGI*direct* Publications.

The industry standard referred to generally for using ECGAs in non-domestic premises is BS 7967-5: 2010 'Carbon monoxide in dwellings and other premises and the combustion performance of gas-fired appliances. Guide for using electronic portable combustion gas analysers in non-domestic premises for the measurement of carbon monoxide and carbon dioxide levels and the determination of combustion performance'.

Completion

Once all aforementioned checks have been made to the complete satisfaction of the commissioning operative the following will be required:

- all users are instructed in the correct operation of the system and its user controls, light-up and shutdown sequence

- manufacturer's 'user' instructions are left for the responsible person. The 'responsible person' is defined by the GSIUR as the 'owner or occupier'. However, on many large commercial or industrial sites this may not be relevant, it may be more appropriate to identify the works engineer or manager for this purpose

- the responsible person should also be advised concerning matters such as actions to be taken in the event of fault or emergency conditions and advice concerning regular maintenance of the units

- a suitable report will need to be completed and left with the responsible person. This report needs to detail the final setting parameters of the heater/system, including:

  - gas user/site details

  - plant details, e.g. make, model and serial number

  - fuel supply details, e.g. type and supply pressure

  - system operating set levels

  - combustion and emission data

  - electrical data, e.g. nominal supply, overload settings, fuse ratings.

Note: CORGI*direct* provides suitable forms for the correct documentation of checks/test carried out, which gas operatives' can provide to their customers. The forms, 'Plant Commissioning/Servicing Record' (Order Ref: CP15), 'Gas Testing and Purging' (Order Ref: CP16) and 'Gas Installation Safety Report' (Order Ref: CP17) are all detailed in Part 13 – CORGI*direct* Publications.

Handover

Upon completion of all the aforementioned commissioning actions, the commissioning report, together with any manufacturer's installation and maintenance instructions are handed over to the designated responsible person for the site/premises.

It should then be made clear to all concerned that the commissioning process has been completed and responsibility for the plant is passed on to the appropriate personnel.

Servicing, maintenance and fault finding – 10

10 – Servicing, maintenance and fault finding

General

Manufacturer's instructions will provide specific information for the servicing and maintenance of any particular appliance and its associated equipment and therefore, should be followed without deviation unless directed by a specific manufacturer.

In general, most appliances will require servicing at least once each year, typically prior to the commencement of the heating season, to maintain their optimum performance and ensure continued safe operation.

In addition, particularly with older appliances, when carrying out servicing work it may be appropriate to upgrade the appliance in line with current regulations, standards and codes of practice.

Similar to the 'planning' stage of the 'commissioning' process discussed in **Part 6 – Boilers, Part 7 – Overhead radiant heating, Part 8 – Warm air heating systems** – and **Part 9 – Combined heat and power**, a risk assessment will be required to establish the requirements in each specific case. In particular, make sure that adequate access arrangements have been made for safe access to the appliances.

All tools and servicing equipment will need to be of sound quality and suitable for the tasks required. Electronic gauges and instruments should be correctly calibrated and certificated accordingly, where necessary.

Only the correct tools should be used for any particular task, e.g. use correctly sized spanners to tighten or undo nuts and bolts – NOT adjustable grips, which may lead to permanent damage of the component and lead to future difficulties in carrying out effective servicing.

When dismantling components for cleaning, ensure that only the suitable cleaning agents, as recommended by the heater manufacturers, are used.

Boilers

The process of servicing an appliance will follow a course broadly similar to that outlined in the generic commissioning process in **Part 6 – Boilers – Commissioning**, except that now the tasks required to service the appliance will be included at the activation stage.

Generic pre-servicing tasks are as follows:

- it is usually a good idea to check with the user prior to commencement of any servicing work to establish if they are aware of any problems with the appliance. It may also be good practice to start ('flash') up the appliance prior to commencing work, just to see if there are any obvious problem areas where attention may need to be directed during the course of the servicing work

- visually inspect the appliance to establish;

    - any signs of malfunction, e.g. evidence of spillage, or soot deposits

    - any signs of external damage, e.g. corrosion, or broken components

    - that the appliance has been installed correctly in accordance with manufacturer's instructions and all relevant standards, codes of practice etc.

    - all manufacturer's stated clearances are correct to enable full servicing, e.g. burner withdrawal

    - the appliance has been correctly assembled and supported

    - gas connections appear to be of the correct size and materials. If the heater has been fitted with a gas flexible connection does it meet the requirements of the current British Standard, e.g. material complying with BS 6501?

Older appliances may be fitted with flexible connections that were supplied by the manufacturer at the time of installation, but do not meet current standards, e.g. a rubber flexible connection with stainless steel over braiding

Generic service tasks are as follows:

- isolate gas and electrical supplies

- dismantle burner assembly, clean the burner and any injector. Check for damage, e.g. heat erosion and rectify problem. Renew burner (natural draught) or burner head/diffuser (forced draught) with new if required. Also check pilot burners on boilers with thermo-electric flame safety for damage. Clean or replace as necessary

- check the burner fan assembly on forced/induced draught appliances. Ensure the fan moves freely, no obvious signs of bearing damage. Clean the fan impellers as necessary

- on boilers with horizontally mounted burners located at the front of the boiler (e.g. forced draught) with the burner assembly removed, it will normally be possible to inspect the heat exchanger interior. On boilers with under-firing burners (e.g. atmospheric) without integral draught diverters it will usually be necessary to also disconnect the flue connector to inspect the heat exchanger(s)

If there is any evidence of carbon deposits, the heat exchanger should be thoroughly cleaned. Soot deposits represent some form of firing malfunction, investigate the cause and rectify before re-commissioning the heater

- inspect the heat exchanger closely for cracks or signs of damage due to water leakage. Establish cause (e.g. flame impingement) and rectify problem. Replace any cast iron heat exchanger section as necessary

If the boiler is outside its expected life span, it could be financially more economical to replace it with a new, more efficient model or system

- check the operation of any air pressure switch. Inspect connecting tubes for blockages or damage etc. Replace as necessary

- check the operation and condition of any flue damper mechanism e.g. on modular boiler systems

- examine the spark ignition device. Clean, reposition for correct spark gap, replace as necessary. Check hard wiring connections replace/reconnect as necessary

- examine the flame detection sensor (UV cell/rectification probe/thermocouple). Clean, reposition, replace as necessary

- check the condense trap if fitted. Ensure that there are no blockages in condense line. Check any non-return valve in the condense line for correct operation. Look for signs of corrosion and rectify

- inspect the chimney/flue pipe for signs of water ingress. Repair/reseal as necessary

- storage water heaters are sometimes fitted with sacrificial magnesium anodes; these anodes will need to be checked for condition and replaced if necessary.

Some manufacturers recommend annual replacement as a matter of course. If the anode is in satisfactory condition but encrusted in lime-scale, the lime-scale will need to be carefully removed and the bare metal of the anode exposed

- some storage hot water units will need to be drained down and inspected internally for signs of scaling.

Where signs of scaling are evident, the manufacturer will need to be consulted on the best course of action.

This will usually entail refilling the system and flushing with the manufacturer's approved de-scaling agent. However, extreme caution should be observed if the hot water from the heater is being used in a food production cycle; any de-scaling agent must be thoroughly flushed through so that it does not contaminate the food product

- any fitted water softening facility will need to be checked to ensure correct operation

- check the water system pressure on pressurised circulating systems, ensure there are no significant water leaks

- large boiler plant such as steel shell boilers on MTHW, HTHW and steam systems will usually need to be inspected annually by the insurance agent.

  It is usually an opportune time to carry out this inspection at the time of servicing, some communication with the insurers agent will be required therefore, to ensure this takes place at a convenient time for all concerned

- reassemble all boiler components. Turn on electricity. With the gas supply isolated, allow the boiler to cycle through the ignition process. Observe the correct sequence and confirm burner lockout

- carry out tightness test on the gas pipework connector

- carry out tightness test of the safety shut off valves or thermo-electric valve; ensure there is no let-by

- re-commission the appliance in accordance with the manufacturer's instructions, also see **Part 6 – Boilers – Commissioning**.

Luminous radiant plaque/cone and overhead radiant heaters

The process of servicing an appliance will follow a course broadly similar to that outlined in the generic commissioning process in **Part 7 – Overhead radiant heating**, except that now the tasks required to service the appliance will be included at the activation stage.

Luminous radiant plaque/cone heaters

Generic pre-servicing tasks are as follows:

- it is usually a good idea to check with the user prior to commencement of any servicing work to establish if they are aware of any problems with the appliance. It may also be good practice to start ('flash') up the appliance prior to commencing work, just to see if there are any obvious problem areas where attention may need to be directed during the course of the servicing work

- visually inspect the appliance to establish:

  - any signs of malfunction e.g. evidence of spillage, or soot deposits

  - any signs of damage e.g. corrosion, broken components, loose or displaced reflectors, damaged tubes etc

  - that the appliance has been installed correctly in accordance with manufacturer's instructions and all relevant standards, codes of practice etc.

  - all manufacturer's stated clearances are correct to enable full servicing e.g. burner withdrawal

  - the appliance has been correctly assembled and supported

- gas connections appear to be of the correct size and materials, particularly with respect to the flexible gas connection.

  If the heater unit is an older model, it may have been installed with a flexible connection provided by the manufacturers at the time of installation, which no longer meets the requirements of the current British Standard, e.g. material to BS 6501, corrugated stainless steel.

Generic service tasks are as follows:

- isolate the gas and electrical supplies

- inspect the radiant surface/ceramic plaque closely for cracks or signs of damage. Replace as necessary.

  Dust from the atmosphere will build-up on the rear face of the ceramic plaque causing localised overheating and light back. Clean the ceramic plaque by blowing compressed air (nominal pressure of 5 bar – check with appliance manufacturers) across the surface – DO NOT attempt to clean the plaque using any form of brush, as this may damage the plaque. Avoid also directing the compressed air at any gasket or sealing material

- clean the burner air/gas mixing venturi and injector

- clean reflectors if necessary

- examine the pilot injector and burner on manual or thermo-electric units. Clean or replace as necessary

- it may be prudent to replace any thermocouple at this stage as a matter of course to avoid nuisance call backs, particularly where access is at high level and erection of scaffolding is required

- examine spark ignition device. Clean, reposition for correct spark gap, replace as necessary

- examine flame detection probe on automatic units. Clean, reposition, replace as necessary

- reassemble all heater components. Turn on electricity to automatic units. With the gas supply isolated, allow the heater unit to cycle through the ignition process. Observe the correct sequence and confirm burner lockout

- carry out a gas tightness test on the gas pipework connector

- re-commission the appliance following the procedure described in the previous section on 'commissioning'.

Overhead radiant tube heaters

The generic pre-servicing tasks are the same as discussed for **Luminous radiant plaque/cone heaters**, but there are some additional requirements for overhead radiant tube heaters, these being:

Generic service tasks

- isolate gas and electrical supplies

- clean reflectors if necessary

- inspect the radiant tube closely for cracks or signs of damage. Establish cause, e.g. flame impingement and rectify problem. Replace tube section as necessary

- lightly brush tube exterior

- dismantle burner assembly, clean the burner and injector. Check for damage, e.g. heat erosion and rectify problem. Replace burner head with new if required

- with the burner assembly removed, it will normally be possible to inspect the tube interior. If there is any evidence of carbon deposits the tube should be lowered to the ground and thoroughly cleaned.

  Soot deposits represent some form of firing malfunction, investigate the cause and rectify before re-commissioning of the heater

- check the operation of the air pressure switch. Inspect connecting tubes for damage, blockages etc., replace as necessary

- examine the spark ignition device. Clean, reposition for correct spark gap, replace as necessary

- examine the flame detection probe. Clean, reposition, replace as necessary

- check the fan assembly. Ensure the fan moves freely, no obvious signs of bearing damage. Clean the fan impellers as necessary

- check the condense trap if fitted. Ensure that there are no blockages in condense line. Check the non-return valve in condense line for correct operation. Look for signs of corrosion and rectify

- reassemble all heater components. Turn on electricity. With the gas supply isolated, allow the heater unit to cycle through the ignition process. Observe the correct sequence and confirm burner lockout

- carry out tightness test on the gas pipework connector

- re-commission the appliance following the procedure described in the manufacturer's instructions. See also **Part 7 – Overhead radiant heating – Commissioning** for further guidance.

Warm air heaters

The process of servicing an appliance will follow a course broadly similar to that outlined in the generic commissioning process in **Part 8 – Warm air heating systems**, except that now the tasks required to service the appliance will be included at the activation stage.

Generic pre-servicing tasks are as follows:

- it is usually a good idea to check with the user prior to commencement of any servicing work to establish if they are aware of any problems with the appliance. It may also be good practice to start-up ('flash') the appliance prior to commencing work, just to see if there are any obvious problem areas where attention may need to be directed during the course of the servicing work

- visually inspect the appliance to establish:

  - any signs of malfunction e.g. evidence of spillage, soot deposits

  - any signs of external damage e.g. corrosion, broken components, loose or displaced louvres etc.

  - that the appliance has been installed correctly in accordance with manufacturer's instructions and all relevant standards, codes of practice etc.

  - all manufacturer's stated clearances are correct to enable full servicing e.g. burner withdrawal

  - the appliance has been correctly assembled and supported

  - gas connections appear to be of the correct size and materials. If the heater has been fitted with a gas flexible connection, does it meet the requirements of the current British Standard, e.g. material to BS 6501, corrugated stainless steel?

Older appliances may be fitted with flexible connections that were supplied by the manufacturer at the time of installation, but do not meet current standards, e.g. a rubber flex with stainless steel over braiding

Generic service tasks are as follows:

- isolate gas and electrical supplies

- clean louvres if necessary. Adjust angle of deflection as necessary

- lightly brush heat exchanger exterior

- lightly brush air distribution propeller fans or remove fan casing on centrifugal fanned units to access and clean fan impellers

- on centrifugal fanned units check condition of fan belts. Replace any damaged belts and adjust for correct tension

Note: If there are multiple fan belts; if only one is damaged, all should be replaced.

- dismantle burner assembly, clean the burner and injector. Check for damage, e.g. heat erosion and rectify problem. Replace burner (natural draught) or burner head/diffuser (forced draught) with new if required. Also check pilot burners on heaters with thermo-electric flame supervision devices for damage. Clean or replace as necessary

- with the burner assembly removed, it will normally be possible to inspect the heat exchanger interior. If there is any evidence of carbon deposits the heat exchanger should be thoroughly cleaned.

  Soot deposits represent some form of firing malfunction, investigate the cause and rectify before re-commissioning of the heater

- inspect the heat exchanger closely for cracks or signs of damage. Establish cause, e.g. flame impingement and rectify problem. Replace heat exchanger section as necessary

- check the operation of any air pressure switch. Inspect connecting tubes for blockages or damage etc. Replace as necessary

- examine the spark ignition device. Clean, reposition for correct spark gap, replace as necessary. Check hard wiring connections replace/reconnect as necessary

- examine the flame detection sensor (UV cell/rectification probe/thermocouple). Clean, reposition, replace as necessary

- check the burner fan assembly on forced/induced draught appliances. Ensure the fan moves freely, no obvious signs of bearing damage. Clean the fan impellers as necessary

- check the condense trap if fitted. Ensure that there are no blockages in the condense line. Check any non-return valve in the condense line for correct operation. Look for signs of corrosion and rectify

- inspect the chimney/flue pipe for signs of water ingress. Repair/reseal as necessary

- reassemble all heater components. Turn on electricity. With the gas supply isolated, allow the heater unit to cycle through the ignition process. Observe the correct sequence and confirm burner lockout

- carry out tightness test on the gas pipework flexible connector, typically achieved using non-corrosive leak detection fluid (LDF)

- carry out tightness test of the safety shut off valves or thermo-electric valve ensure there is no let-by

- re-commission the appliance following the procedure described in the previous section on 'commissioning'.

Combined heat and power

The process of servicing an appliance will follow a course broadly similar to that outlined in the generic commissioning process of **Part 9 – Combined heat and power**, except that now the tasks required to service the appliance will be included at the activation stage.

The CHP control unit can usually provide data based on the performance of the engine and can provide prescriptive action to follow during the servicing period.

Generic pre-servicing tasks are as follows:

* it is usually a good idea to check with the gas user prior to commencement of any servicing work to establish if they are aware of any problems with the CHP unit. It may also be good practice to start ('flash') up the engine prior to commencing work, just to see if there are any obvious problem areas where attention may need to be directed during the course of the servicing work

* visually inspect the appliance to establish:

    - any signs of malfunction e.g. evidence of spillage, soot deposits

    - any signs of external damage e.g. corrosion, broken components

    - that the unit has been installed correctly in accordance with manufacturer's instructions and all relevant standards, codes of practice etc

    - all manufacturer's stated clearances are correct to enable full servicing

    - the appliance has been correctly assembled and supported

    - gas connections appear to be of the correct size and materials. If the CHP unit has been fitted with a gas flexible connection does it meet the requirements of the current British Standard, e.g. material complying with BS 6501?

    - examine the condition of the gas flexible connection for signs of deterioration, replace as necessary.

Generic intermediate service tasks may include the following:

* isolate gas and electrical supplies

* visually inspect all gas components, particularly gas flexible connections for signs of wear or damage and replace as necessary

* replace lubricating oil and fill to required level. Dispose of old oil at waste oil collection centres

* replace oil filter

* lubricate all bearings, e.g. generator bearings

* inspect air intake system and replace air filter

* inspect the engine cylinder head valve clearance and adjust to manufacturers required clearances if necessary

* if there is any evidence of carbon deposits at the exhaust outlet it should be thoroughly cleaned.

    Soot deposits represent some form of firing malfunction, investigate the cause and rectify before re-commissioning of the engine

* inspect the heat exchanger(s) closely for cracks or signs of damage due to water leakage. Establish cause and rectify problem

* check the operation of any air or gas pressure switches. Inspect connecting tubes for blockages or damage etc. Replace as necessary

- check the operation and condition of any flue damper mechanism e.g. when interconnected with modular boiler systems

- examine the spark ignition device. Clean, reposition for correct spark gap, replace as necessary. Check hard wiring connections replace/reconnect as necessary

- check the condense trap if fitted. Ensure that there are no blockages in condense line. Check any non-return valve in the condense line for correct operation. Look for signs of corrosion and rectify

- inspect the exhaust system for signs of water ingress. Repair/reseal as necessary

- reassemble all components. Turn on electricity

- with the gas supply isolated, carry out tightness test on the gas pipework connector

- carry out tightness test of the SSOV; ensure there is no let-by

- turn on gas and re-commission the CHP unit following the procedure described in the previous Part on 'commissioning'.

In addition to intermediate servicing tasks generic full service tasks may include the following:

- replacing spark plugs and spark plug cap

- replace ignition coils

- replacing heating water hoses

- replace control system back up batteries

- any other work as specified by the manufacturer.

Once all servicing work has been completed, document and report to the responsible person all work undertaken. If the work is undertaken in association with the CHP unit manufacturer, the manufacturer will also need a copy of the report.

Fault finding

From time to time, appliances will present faults that require investigation by a suitably competent operative to identify, locate and rectify the fault in order to put the appliance back in to operation.

Fault finding should not be difficult when approached in a methodical manner, using information provided by the appliance/equipment manufacturer (typically this information is provided in the form of flow charts and tables, which will give advise to common faults encountered), coupled with knowledge and understanding of the operative of the appliance and using the correct tools and instrumentation.

Where a fault cannot be traced or is not adequately catered for within the manufacturer's literature, contact that manufacturer direct for further guidance.

Always ensure that any fault finding work is carried out in a safe manner and typically after a thorough risk assessment has been carried out of the dangers present such as confined spaces; excessive heat; entanglement; electric shock risk; working at height; vehicular access etc. and that any isolation/lock-off procedures are followed without deviation.

For further information on risk assessments see Essential Gas Safety – Non-Domestic – Part 3 – Risk assessments (Order Ref: ND1), see **Part 13 – CORGI*direct* Publications**).

The person(s) responsible for the appliance(s)/equipment should be kept informed of your work and suitable warnings should be affixed to inform passers-by of the work being carried out.

The following fault finding tables are offered for information purposes only; for a given appliance, reference should be made to the appliance manufacturer's instructions.

Table 10.1 Boilers – atmospheric burner systems with thermo-electric FSD

Symptom	Possible cause	Remedy
Pilot/start gas will not light	gas not turned on at meter or appliance	turn gas on at meter or appliance
	air in gas line	ensure gas is available at the appliance, purge the gas connector
	incorrect pilot lighting procedure	allow thermoelectric valve to 'click'. Follow correct lighting instructions
	pilot injector blocked	clean/replace pilot injector
		check lint filter, if fitted
	no spark at igniter	check spark gap is correct
		check electrical connections to igniter are satisfactory
		check the spark is not tracking to earth, make sure lead is routed clear of metal
		check/replace spark generator
Pilot lights but will not remain alight	thermocouple connections loose	tighten thermocouple connections
	thermocouple damaged	fit new thermocouple
	pilot flame not sufficient to heat thermocouple	pilot injector partially restricted. Clean
		insufficient gas pressure, check available pressure is correct
	faulty thermoelectric valve	check flame signal strength
		replace thermoelectric valve
	interrupted thermocouple activated (open circuit)	investigate cause, e.g. overheat, rectify and reset
Pilot established but main gas will not light	gas not turned on at burner	turn gas on
	gas pressure too low	check supply pressure is ok
	no electricity to control valve	check electricity turned on
		check all controls calling for heat and are working ok
		check all fuses and replace if necessary after further investigation as to the cause
		check wiring for faults/breaks, repair/replace as necessary
	gas solenoid valve stuck	replace valve
	gas regulator stuck closed	replace regulator

Table 10.1 Boilers – atmospheric burner systems with thermo-electric FSD (continued)

Symptom	Possible cause	Remedy
Main burner established but only on low flame	gas pressure too low	adjust appliance regulator to manufacturers data
		check supply pressure:
		– if supply pressure ok, check pipe size. Replace undersized pipe
		– if pipe size ok, check for restrictions (kinked/damaged pipe), replace as necessary
		inlet gauze to multifunctional valve linted. Clean
	gas regulator not working correctly	check vent hole in regulator is not blocked. Clean
		replace regulator
Main burner established but burner has yellow flame	gas rate too high	check/reset gas pressure
	gas injector/filter linted or blocked	clean injector and/or filter
	incorrect burner jets fitted	replace jets
	insufficient ventilation	check ventilation requirement. Fit additional vents as required
	recirculation of combustion products	check for spillage with smoke match. Check exhaust and ventilation is adequate
Main burner established but lightback occurring. Popping noise from heater	broken or cracked plaque	replace damaged plaque
	burner seal damaged	replace seal
	gas supply pressure too high	check/reset gas pressure

Table 10.2 Boilers – atmospheric burner systems with automatic ignition

Symptom	Possible cause	Remedy
Boiler will not light. Control box in shutdown condition	no electrical supply	check electricity turned on
		check all controls calling for heat and are working ok
		check all fuses and replace if necessary
		check wiring for faults/breaks repair/replace as necessary
Boiler will not light. Control box in lockout condition	unknown	observe ignition sequence to determine point at which lockout occurs. Determine fault from manufacturer's diagnostic chart(s). Rectify
	ignition failure	check spark electrode. Clean, reposition, replace as necessary
		check HT lead. Replace as necessary
		check spark generator. Replace as necessary
Burner lights but goes to lockout after 3 seconds	inadequate flame signal	check electrode position. Adjust as necessary
		broken electrode. Replace
		incorrect gas pressure, flame lifting off. Reset to manufacturers data
		broken/disconnected wiring. Check and replace as necessary
		faulty control box. Replace
Burner lights but spark continues after flame established	faulty control box	replace control box
	incorrect wiring. Phase inversion	check wiring against manufacturer's data. Rectify as necessary

Table 10 3 Boilers – forced draught burner systems

Symptom	Possible cause	Remedy
Burner will not start	electrical supply failure	check main isolator switch is on and power is available at the control panel
		check overload protection device of burner motors (if fitted)
	air pressure switch on burner not operating	switch contacts must be in the burner start position
		Check switch is not stuck in the satisfied position
	control thermostat, time, water pressure switches or other boiler system controls at fault	check that all switches are closed
		check booster is not locked out due to low inlet gas pressure and that pressure switches function correctly
		reset and check. Also check control thermostat setting and operation
Burner starts, i.e. burner motor runs but will not light and burner control box goes to 'lockout'	no ignition spark	check condition and position of spark probe
		check condition of HT lead
	gas starvation	purge gas pipe, check pilot valve operation
	ultra-violet cell fault	remove, clean and check connections and replace if necessary
Burner motor starts but stops and recycles	air pressure switch on the burner	check burner damper is open to usual extent
		check air inlet to switch is not blocked
		check switch is not stuck
		check electrical connection
Burner shuts down during operation and recycles	air pressure switch on the burner	check burner damper is open to usual extent
		check air inlet to switch is not blocked
		check switch is not stuck
		check electrical connection
Burner shuts down during operation and goes to 'lockout'	flame failure	check gas supply and pressure
	ultra-violet cell fault	remove, clean and check connections and replace if necessary

Table 10.3 Boilers – forced draught burner systems (continued)

Symptom	Possible cause	Remedy
Burner lights on pilot, but burner locks out after 3-4 seconds	ultra-violet cell faulty	remove, clean and check connections and replace if necessary
	ultra-violet cell is not receiving a sufficient signal from the flame	clean cell and check correctly inserted, facing the flame
Smell of gas local to the appliance	leakage from pipe line joints or pilot or that the main gas valves are not gas tight	switch OFF burner and check pipe line joints with non-corrosive leak detection fluid (LDF)
		close pilot manual valve and then main manual valve at the burner to see if this isolates the problem
		check the automatic valves. Do not operate until this has been done
Combustion noise becomes louder, flame becomes yellow Boiler door becomes hotter than usual	air inlet accidentally masked	clear obstruction
	air inlet damper position moved	reset and make a combustion check
	burner head parts have moved, become worn or overheated	adjust or replace and make combustion check
	gas pressure alteration	check and find cause
	boiler tubes, flueways or chimney blocked or restricted	check and clear restriction
	door refractory failed	replace if damaged is severe

Table 10.4 Overhead radiant heaters – luminous radiant plaque/cone heater

Symptom	Possible cause	Remedy
Pilot/start gas will not light	gas not turned on at meter or appliance	turn gas on at meter or appliance
	air in gas line	ensure gas is available at the appliance, purge the gas connector
	incorrect pilot lighting procedure	allow thermoelectric valve to 'click'. Follow correct lighting procedure within the instructions
	pilot injector blocked	clean/replace pilot injector
		clean in-line filter if fitted
	no spark at igniter	check spark gas is correct
		check electrical connections to igniter are satisfactory
		check the spark is not tracking to earth, make sure lead is routed clear of metal
		check/replace spark generator
Pilot lights but will not remain alight	thermocouple connections loose	tighten thermocouple connections
	thermocouple damaged	fit new thermocouple
	pilot flame not sufficient to heat thermocouple	pilot injector partially restricted. Clean
		insufficient gas pressure, check available pressure is correct
	faulty thermoelectric valve	check flame signal strength
		replace thermoelectric valve
Pilot established but main gas will not light	gas not turned on at burner	turn gas on
	gas pressure too low	check supply pressure is ok
	no electricity to control valve	check electricity is turned on
		check all controls calling for heat and are working ok
		check all fuses and replace if necessary after further investigation as to the cause
		check wiring for faults/breaks, repair/replace as necessary
	gas solenoid valve stuck	replace valve
	gas regulator stuck closed	replace regulator

Table 10.4 Overhead radiant heaters – luminous radiant plaque/cone heater (continued)

Symptom	Possible cause	Remedy
Main burner established but radiant surface/plaques do not glow	gas pressure too low	adjust appliance regulator to manufacturer's data
		check supply pressure is ok
		if supply pressure ok, check pipe size. Replace undersized pipework
		if pipe size ok, check for restrictions (kinked/damaged pipework)
		inlet gauze to multifunctional valve blocked. Clean
	gas regulator not working correctly	check vent hole in regulator is not blocked. Clean
		replace regulator
Main burner established but radiant surface/plaques emits a yellow flame	gas rate too high	check/reset gas pressure
	gas injector/filter linted or blocked	clean injector and/or filter
	incorrect burner jets fitted	replace jets
	insufficient ventilation	check ventilation requirement. Fit additional vents, where required
	recirculation of combustion products	check for spillage with smoke match. Check exhaust and ventilation is adequate
Main burner established but lightback occurring. Popping noise from heater	broken or cracked plaque	replace damaged plaque
	burner seal damaged	replace seal
	gas pressure too high	check/reset gas pressure

Table 10.5 Overhead radiant heaters – atmospheric burner systems with automatic ignition

Symptom	Possible cause	Remedy
Heater will not light. Control box in shutdown condition	no electrical supply	check electricity turned on
		check all controls calling for heat and are working ok
		check all fuses and replace if necessary after further investigation as to cause
		check wiring for faults/breaks, repair/replace as necessary
Heater will not light. Control box in lockout condition	unknown	observe ignition sequence to determine the point at which lockout occurs. Determine fault from manufacturer's diagnostic charts. Rectify fault
	ignition failure	check spark electrode. Clean, reposition, replace as necessary
		check HT lead. Replace as necessary
		check spark generator. Replace as necessary
Burner lights but goes to lockout after 3 seconds	inadequate flame signal	check electrode position. Adjust as necessary
		broken electrode. Replace
		incorrect gas pressure, flame lifting off. Reset to manufacturer's data
		broken/disconnected wiring. Replace
		faulty control box. Replace
Burner lights, but spark continues after flame established	faulty control box	replace control box
	incorrect wiring. Phase inversion	check wiring against manufacturer's data. Rectify as necessary

Table 10.6 Overhead radiant tube heaters

Symptom	Possible cause	Remedy
Burner will not start, 'mains on' does not illuminate.	external controls, thermostats, time switch etc. not calling for heat	adjust controls
	fuse blown	check for short circuit in wiring or individual electrical components
Light illuminates	fan's connector plug not fully engaged	engage securely
	loose electrical connection	check all connections
	fan seized or faulty motor	replace fan, re-commission heater, check gas pressure settings
	sequence controller relay failing to pull in and/or hold in	check vacuum switch is satisfactory, replace sequence controller
	vacuum switch not returning to normal (switch off) position	replace vacuum switch
HB Herringbone system only	fan tripped out on overload	check overload setting
		rest button on in control panel
		check for build-up of condensate in fan casing
Fan starts but burner does not attempt ignition	insufficient vacuum generated by fan	clean fan blades with soft bristle
	blocked emitter tube	clean emitter tube internally
	combustion chamber cover permitting air leakage	examine condition of sealing gasket, tighten down lid securely
	vacuum impulse line between combustion chamber and vacuum switch insecure or defective	fix securely in place
	vacuum switch 'pulls in' but electronic sequence controller does not proceed to programmer ignition sequence	replace burner sequence controller unit, but first check that the cause of failure is not a short on output circuit
Herringbone system only	leaks in joint, separation in manifold system	reseal or reconnect

Table 10.6 Overhead radiant tube heaters (continued)

Symptom	Possible cause	Remedy
Burner proceeds to ignition stage (normally indicated by audible spark valve energised and 'burner on' light illuminated) but burner does not light.	no spark	check electrode for cracks – replace if necessary
		check high tension connections are secure
		check spark gap. If no high tension output from electronic controller, replace controller
	gas safety valve faulty or defective	replace solenoid operator section of gas valve
	insufficient gas pressure	set burner pressure to that indicated on data badge
Burner lights but shuts down after a few seconds	inadequate flame signal	check/replace electrode
	required signal in accordance with manufacturer's instructions	check connections to electrodes and electric sequence controller plug
		replace the electronic sequence controller
	flame unstable	check cleanliness of burner and set burner pressure as indicated on data badge
	inadequate gas supply. Observe burner gas pressure, with all heaters operating	if gas pressure drops below that indicated on the data badge, examine gas supply pipework for excessive pressure loss
	insufficient vacuum at combustion chamber causing vacuum switch to cut off	clean fan blades with soft bristle brush
		inspect tube internally and clean if necessary (see servicing instructions)
Heater shuts down after operating for a period of time	refer to above	if problem persists, replace vacuum switch

Table 10.7 Warm air heating systems – Indirect-fired air heaters – atmospheric burner system with thermo-electric FSD

Symptom	Possible cause	Remedy
Pilot will not light	electrical – no spark	check ignition lead connections
		check spark electrode gap
		faulty piezo igniter – change
		faulty control box – change (auto ignition units)
	gas	check gas supply to pilot assembly
		no pilot flame
		start button on multifunctional control not depressed for long enough
Pilot lights but goes out when start button released	electrical	faulty thermocouple – check connections – change
		higher limit thermostat tripped out – check for reason, rectify and reset
		faulty multifunctional control – change
	gas	pilot flame too small – check and adjust as necessary
Main burner will not light	electrical	check electrical supply is ON
		check controls are ON or calling for heat
		faulty multifunctional control – change
		no ignition spark or rectification signal (auto ignition units)
		control box at lockout – reset, or faulty – change (auto ignition units)
Main burner lights, but goes out before main fan comes on	electrical – unit goes out on high limit	check fan command module (timer) operation
		check limit thermostat setting – relight pilot
		faulty limit thermostat – change and relight pilot
		faulty fan assembly – change
		ducted units only – fan motor out on thermal overload. Allow motor to cool then check running amps
		check duct resistance

Table 10.7 Warm air heating systems – Indirect-fired air heaters – atmospheric burner system with thermo-electric FSD (continued)

Symptom	Possible cause	Remedy
Main fan runs continuously.	electrical	summer/winter switch and/or external controls set to summer
		fan command module (timer) faulty – change
Main fan fails to run.	electrical	fan motor or capacitor failed – replace
		fan command module (timer) faulty – change
		fan contactor failed – replace

Table 10.8 Warm air heating systems – Indirect-fired air heaters – forced draught burner systems

Symptom	Possible cause	Remedy
Burner will not light.	no electricity	check fused isolator
		check switched on
		check control panel
		check fuses
		check supply voltage
	electricity confirmed – burner still not lighting	check overheat reset
		check burner
		check lockout
		check flame probe
		check air pressure switch
		check gas pressure switch
		check stat setting
		check time clock
		check burner motor
Burner starts but goes to lockout.	check gas supply	main gas cock open
		gas train gas cock open
		gas pilot valve open
	air in gas line	purge gas line

Table 10.8 Warm air heating systems – Indirect-fired air heaters – forced draught burner systems (continued)

Symptom	Possible cause	Remedy
Burner starts but goes to lockout	flame probe failure	faulty probe – replace
		bad earth – check earth connections
		wrong position – check against instructions
		faulty control box – replace control box
	ignition failure	faulty control box – replace control box
		faulty ignition transformer – check HT lead
		faulty electrode – check gap, adjust or replace
	air pressure switch failure	check settings against instructions, replace if necessary
Burner lights but will not hold on main flame	gas pressure	check head pressure as per instructions
		check inlet pressure
	excess combustion air; or combustion air too low	check setting of air flap
		check CO_2 level at flue
		check head setting
		adjust air flap
		adjust heat setting
	check flame probe	position
		signal
		check main gas actuator
	check control box	check main gas actuator
Burner lights but will not hold on main flame	insufficient air	air inlets blocked – remove obstruction and clean filters
		outlet nozzles closed – open louvres on nozzles
		duct resistance too high – open duct grilles
		fan belts slipping – adjust fan belts
	high inlet pressure	warm air recirculation – adjust nozzle position

Table 10.8 Warm air heating systems – Indirect-fired air heaters – forced draught burner systems (continued)

Symptom	Possible cause	Remedy
Burner lights but will not hold on main flame	fan limit stat	replace
		incorrect settings – reset
	incorrect gas pressure	mains pressure – adjust settings to instructions
		head pressure – adjust settings to instructions

Table 10.9 Warm air heating systems – direct-fired air heaters

Symptom	Possible cause	Remedy
Heater will not start	no power supply	turn on main isolator
	fuse blown	find fault. Replace fuse
	fan on overload	reset overload
	heater locked out	reset lockout
	overheat thermostat activated	manually reset and check setting is 80°C
	micro-processor not programmed	set up programmer
Programmer cycles continuously	air pressure switch not acting, possibly welded	reset pressure or replace faulty switch
Programmer locks out pilot does not light	no gas	turn on gas. Check pilot solenoid is opening
	no spark	check spark lead and gap
	insufficient airflow	belts slipping. Duct blocked, filter blocked
Programmer locks out, but burner does light	flame not seen by sensor	check scanner leads, replace faulty scanner
	for box burner flame probe not sensing flame	check probe leads, cap, and replace faulty flame probe
	flame signal not accepted by programmer	check programmer. Replace programmer
Programmer locks out when flame is on.	unstable main flame	check fan belts for slipping
	lack of gas	check gas supply

Combined heat and power

When faults arise on CHP systems the control unit can usually be investigated to establish diagnostic data on which to base appropriate corrective action.

The appropriate corrective work will usually need to be carried out in accordance with the particular manufacturer's written instructions or verbal advice.

Any replacement components will need to be of sound quality and approved by the manufacturer for the specific application.

The following are some typical faults that can occur.

Table 10.10 Combined heat and power

Symptom	Possible cause	Remedy
Coolant temperature too low	ambient temperature too low – too much cooling air intake	check air intake system shutters for correct operation
	temperature gauge malfunction	check and replace gauge as necessary
	engine enclosure thermostat inoperative	check and replace thermostat as necessary
Engine coolant temperature too high	temperature gauge malfunction	check and replace gauge as necessary
	radiator damaged	repair/replace radiator as necessary
	radiator cap damaged or incorrect	replace radiator cap with correct type
	cooling system hoses restricted or leaking	replace as necessary
Poor engine acceleration/ response or: engine power output low; engine runs rough or misfires; engine starts but will not keep running; engine will not reach rated speed	fuel leak	check engine fuel line with LDF and repair
	air intake or exhaust leak	check air and exhaust pipes and repair
	air intake restriction	inspect air intake and clean. Replace filter
	exhaust restriction	inspect and clean
	fuel intake incorrect/restricted	check fuel throughput and rectify
	ignition system misfiring	check and clean spark plug gaps. Replace spark plugs
Engine will not start	gas not turned on at meter or engine	turn gas on at meter or engine
	air in gas line	ensure gas is available at the engine, purge the gas connector

Table 10.10 Combined heat and power (continued)

Symptom	Possible cause	Remedy
Engine will not start	no spark at igniter	check spark gap is correct
		check electrical connections to igniter are satisfactory
		check the spark is not tracking to earth, make sure lead is routed clear of metal
		check/replace spark generator
	air intake restricted	inspect and clear restriction
Engine will not stop	engine running on fumes drawn into air intake	check for source of fumes and rectify
		check location of air intake

11 – Definitions

Definitions

1st family gas: at present normally only LPG-air.

2nd family gas: Natural gases.

3rd family gas: liquefied petroleum gases (LPG).

Aerated burner: a burner in which some or all of the required air has been mixed with the gas before it leaves the burner port(s).

Air heater: appliance for heating air to be used for space heating.

Air vent: non-adjustable grille or duct which allows the passage of air at all times.

Air vent free area: total area of the unobstructed openings of an air vent.

Appliance manufacturer's installation instructions: the instructions prepared by the appliance manufacturer giving particular information and requirements on how the appliance should be installed.

Appliance ventilation duct: provided to convey combustion or cooling air for an appliance or component.

Appliance, gas: appliance designed for heating, lighting, cooking or other purposes.

Balanced flued appliance: a room-sealed appliance which draws its combustion air from a point adjacent to that at which the combustion products are discharged, the inlet and outlet being so disposed that wind effects are substantially balanced.

Basement: a room which is completely or partly below ground level on all or some sides. (LPG appliances).

Boiler: appliance designed to heat water for space heating and/or water supply.

Branched flue system: a shared open-flued system serving appliances situated on two or more floors.

Calorific Value (or CV): the Calorific Value is the quantity of heat (energy) produced when a unit volume of the fuel, measured under standard conditions of temperature and pressure, is burned completely in excess air.

A distinction is made between:

a) Gross Calorific Value – (also known as the Higher Calorific Value or HCV) – in the determination of which the water vapour produced by the combustion of the fuel is deemed to have been condensed into a liquid phase at the standard temperature and its latent heat released.

b) Net Calorific Value – (also known as the Lower Calorific Value or LCV) – in the determination of which water vapour produced by the combustion of the fuel is assumed to remain in the vapour phase. The net calorific value is therefore the gross calorific value minus the latent heat of the water vapour contained in the combustion products.

Chimney: a structure consisting of a wall or walls enclosing a flue (see **Flue** in this Part) or number of flues - the term applies equally to masonry, metallic, plastic, etc. systems used for both open-flued and room-sealed appliances.

Chimney component: any part of a chimney.

Commissioning: initial start-up of an installation to check and adjust for safe and reliable operation.

Common flue system: a shared open-flued system that serves two or more appliances installed in the same room or space.

Compartment: housing specifically designed or adapted to house a gas appliance (see also **Balanced compartment**).

Competence: competence in safe gas installation requires enough knowledge, practical skill and experience to carry out the job in hand safely, with due regard to good working practice. Knowledge must be kept up to date with awareness of changes in law, technology and safe working practice.

Condensing appliance: designed to use latent heat from water vapour in the combustion products by condensing the water vapour within the appliance.

Condensate drain: a device in a flue where condensate can be removed.

Condensate-free length: the length of individual open-flue, which can be calculated to maintain the temperature of the flue gases above the dew point.

Damper: a device used to vary the volume of air passing through a confined cross-section by varying the effective sectional area.

Data plate: a durable, permanently fixed plate bearing specified information relative to the appliance.

Dry run: verification of the correct operation of components and interlocks without fuel being available to the combustion space.

Ducted warm air heater: flued appliance which uses ducts to distribute the heated air.

Emergency control valve: valve for shutting off the supply of gas in an emergency; not a service valve.

Enforcing authority: authority with a responsibility for enforcing the Health and Safety at Work etc. Act 1974 and other relevant statutory provisions; normally HSE or the local authority for the area as determined by the Health and Safety (Enforcing Authority) Regulations 1977.

Flame supervision device: control which detects the presence of a flame and in the absence of that flame, prevents the uncontrolled release of gas to the burner.

Flexible connector: pipe connector to a bayonet valve that allows a moveable flueless appliance to be safely disconnected; allows an appliance to be moved a short distance without the need for disconnection.

Flue: passage or space for conveying the products of combustion from a gas appliance to the outside air.

Flue break: an opening in the secondary flue in the same room as and in addition to, the opening at the draught diverter.

Flue lining: a wall of a chimney consisting of components the surface of which is in contact with the products of combustion. This includes a rigid or flexible chimney or flue liner inserted into a chimney.

Flue outlet: the part of the appliance that allows the exit of products of combustion from the appliance.

Flue pipe: a pipe enclosing a flue; for a double wall type it is the inner pipe.

Flueless appliance: designed for use without a flue under conditions in which the products of combustion do not present a safety risk.

Free area: total area of the individual unobstructed openings of an air vent.

Gas fittings: gas pipes, regulators and meters and fittings, apparatus and appliances, designed for use by consumers of gas for heating, lighting and other purposes, for which gas can be used (other than the purpose of an industrial process carried out on industrial premises). It does not mean:

a) Any part of a service pipe.

b) Any part of a distribution main or other pipe upstream of the service pipe.

c) A gas storage vessel.

d) A gas cylinder or cartridge designed to be disposed of when empty.

Gas Safety Regulations: legally binding requirements for safe gas work.

Identity badge: see **Data plate**.

Lock-out: a safety shut-down condition of a control system such that restart cannot be accomplished without manual intervention.

Louvre: an assembly of fixed angled blades or vanes contained within a framework, which is designed to resist light and rain penetration but allow the passage of air at all times.

Manufacturer's instructions: instructions provided by the manufacturer pertaining to the correct installation, adjustment and use of equipment, which need to be followed.

Mechanical ventilation: air supplied by a fan.

Natural draught flue system: a flue system in which the draught is provided by the thermal force arising from the heat of the products of combustion.

Open-flued appliance: an appliance designed to be connected to an open-flued system which draws combustion air from the room or space in which it is installed.

Open-flued system: system that is open to a room or internal space at each appliance.

Pluming: visible cloud of combustion products from an outside flue terminal which are cooled to below dew point by mixing with external air.

Pre-aerated burner: a burner to which gas and air are supplied already mixed.

Remotely operated emergency valve: shut-off valve capable of remote operation, which will close automatically on loss of the actuating power or fire engulfment and which, preferably, is fire safe in accordance with BS EN 12266: 2012, parts 1 & 2 (replaces BS 5146). Electrically driven valves need not close automatically on fire engulfment if the power supply is adequately fire protected.

Responsible person: the occupier of the premises or, where there is no occupier or the occupier is away, the owner of the premises or any person with authority to take appropriate action in relation to any gas fitting therein.

RIDDOR: The Reporting of Injuries, Diseases and Dangerous Occurrences Regulations.

Room-sealed: an appliance that, when in operation, has the combustion system, including the air inlet and the products outlet, isolated from the room or space in which the appliance is installed.

Safety shut-off valve: actuated by the safety control so as to admit and stop gas flow automatically.

Secondary flue: the part of the open-flued system connecting a draught diverter or draught break to the terminal.

Semi-rigid stainless steel connector: a stainless steel tube formed with annular corrugations and having factory-fitted end connections.

Swirl plate: a long flat plate of stainless steel that has been twisted about its axis along its entire length to form a helical spiral. When the plate is inserted in the flue path (tubes) of the heat exchanger POCs follow the spiral path created by the swirl plate. In this way the turbulent flow of POCs created maximises heat transfer through the heat exchanger.

Terminal: device fitted at the flue outlet to:

a) Assist the escape of products of combustion.

b) Minimise downdraught.

c) Prevent flue blockages.

Termination: the outlet of a flue system where products of combustion discharge into external air.

Thermal cut-off device: safety device designed to stop the flow of gas when the surrounding air temperature exceeds the predetermined value.

Tightness test: the testing of installation pipes and equipment for escapes from the system.

Valve: device to stop or regulate the flow of gas by the closure or partial closure of an orifice by means of a gate, flap or disc.

Ventilation: the process of supplying fresh air to and removing used air from, a room or internal space.

Work: in relation to a gas fitting this includes any of the following activities carried out by any person, whether an employee or not:

a) Installing the fitting.

b) Maintaining, servicing, permanently adjusting, repairing, altering or renewing the fitting or purging it of air or gas.

c) Changing the position of a fitting when it is not readily removable.

d) Removing the fitting.

12 – References

Manufacturers/suppliers

Advanced Combustion Engineering ltd (industrial burners)

Tel: 01706 212 218
E mail: info@aceburners.co.uk
Website: www.aceburners.co.uk

Aerogen Company ltd (flame treatment specialist)

Tel: 01420 83744
E mail: info@aerogen.co.uk
Website: www.aerogen.co.uk

Aeromatix (a division of Worgas Burners ltd)

Tel: 01773 864 870
E-mail: info@aeromatix.com
Website: www.aeromatix.com

Airflow (Nicoll Ventilators) Ltd

Tel: 01425 611 547
E-mail: sales@airflow-vent.co.uk
Website: www.airflow-vent.co.uk

Airflow (industrial fans etc)

Tel: 01494 525 252
Website: www.airflow.co.uk

Air Products plc (gasses and air)

Tel: 0800 389 0202
Website: www.airproducts.co.uk

Alpha Therm Ltd

Tel: 0844 871 8760
Technical Helpline: 0844 871 8764
Training Academy: 0844 871 8763
Website: www.alpha-innovation.co.uk

Ambaheat (heating & cooling systems)

Tel: 01543 878 772
Website: www.ambaheat.co.uk

Ambi-Rad Ltd

Tel: 01384 489 700
Website: www.ambirad.co.uk

Andrews Water Heaters

Tel: 0845 070 1057
E-mail: andrews@baxicommercialdivision.com
Website: www.andrewswaterheaters.co.uk

Arctic Products ltd (pipe freezing equipment)

Tel: 0844 871 8461
E mail: sales@arctic-products.co.uk
Website: www.arctic-products.co.uk

Ariston Thermo Group (includes Ariston and Elco brands)

Technical Helpline: 0333 240 7777
E-mail: techhelpdesk.uk@aristonthermo.com
Website: www.ariston.com

Barlo Radiators

Tel: 07876 218 816
Website: www.barlo-radiators.com

Baxi Commercial Division

Tel: 0845 070 1055
Website: www.baxicommercial.co.uk

B.E.S. Ltd

FREEFONE: 0800 80 10 90
Website: www.bes.co.uk

Biasi UK Ltd

Tel: 01922 714 600; option 2
E-mail: service@biasi.uk.com
Website: www.biasi.co.uk

Black Teknigas and Electro Controls Ltd (gas control equipment)

Tel 01480 407 074
E mail sales@blackteknigas.co.uk
Website: www.blackteknigas.com

Brewer Metalcraft
(chimney cowls & terminals)

Tel: 0845 676 0702
Website: www.brewercowls.co.uk

BP Gas
(LP gas supplier)

Tel 0845 300 0038
Website: www.bpgas.co.uk

Bryan Donkin USA
(gas control equipment)

Website: www.bryandonkinusa.com

Mantec Technical Ceramics Ltd
(specialist in ceramic applications)

Tel: 01782 377 550
Website: www.mantectechnicalceramics.com

Clyde Energy Solutions Ltd

Tel: 01342 305 550
E-mail: info@clyde4heat.co.uk
Website: www.clyde4heat.co.uk

Colt Group Ltd

Tel: 02392 451 111
E-mail: info@coltgroup.com
Website: www.coltgroup.com

Comark Instruments
(test & monitoring equipment)

Tel: 0844 815 6599
E mail: technical@comarkltd.com
Website: www.comarkltd.com

Danfoss Randall Ltd

Tel: 0845 121 7400
E-mail: danfossrandall@danfos.com
Website: www.danfoss-randall.co.uk

Dräger Safety UK Ltd

Tel: 01670 352 891
Website: www.draeger.co.uk

Ellison Energy Services ltd (Boilers)

Tel: 01274 533 207
E Mail: info@ellisonenergy.co.uk
Website: www.ellisonenergy.co.uk

Elster Metering ltd

Website: www.elstermetering.co.uk

Ferroli Ltd

Tel: 0843 479 0479 (Technical)
E-mail: service@ferroli.co.uk
Website: www.ferroli.co.uk

Federation of Heating Spares Stockists (FHSS)

Tel: 0844 636 0696
Website: www.heat-spares.co.uk

Fine Tubes ltd

Tel 01752 735 851
E mail: feedback@fine-tubes.co.uk
Website: www.finetubes.co.uk

Fiorentini UK ltd
(pressure regulators, meters etc)

Tel 01926 814 866
E mail: sales@fiorentiniuk.com
Website: www.fiorentiniuk.com

Flanges ltd
(flanges, orifices)

Tel 01642 672 626
E mail: enquiries@flanges-ltd.co.uk
Website: www.home.btconnect.com/Flanges-Ltd

Flueboost ltd
(flue boosters)

Tel 01565 755 599
E-mail: flueboost@yahoo.co.uk
Website: www.flueboost.co.uk

Foster Wheeler Energy ltd

Website: www.fwc.com

F&P Wholesale

Tel: 0116 256 7380
E-mail: info@fpwholesale.co.uk
Website: www.fpwholesale.co.uk

Fulton Ltd

Tel: 0117 9723 322
Website: www.fulton.co.uk

Gas & Environmental Services

Tel: 01707 373 751
E-mail: info@gesuk.com
Website: www.gesuk.com

Gas Appliance Distributors
(acquired by Stearn)

Website: www.stearn.co.uk

Gas Appliance Spares (Preston)

Tel: 01772 702 755
E-mail: info@gas-spares.co.uk
Website: www.gas-spares.co.uk

Gledhill Products ltd

Tel 01253 474 550
Website: www.gledhill.net

Glow-Worm

Tel: 01773 828 300 (Technical Helpline)
Website: www.glow-worm.co.uk

Glynwed pipesystems
(comprising of Durapipe Uk, Philmac, GPS & Astore)

Website: www.glynwedpipesystems-uk.com

Grundfos Europump UK

Tel: 01525 850 000
Website: www.europump.co.uk

Halstead (Part of Glen Dimplex Boilers)

Tel: 0844 371 1111
Website: www.glendimplexboilers.com

Hamworthy Heating

Tel 0845 450 2865
Website: www.hamworthy-heating.com

Harton Services Ltd

Tel: 020 8310 0421
E-mail: office@hartonservices.com
Website: www.hartons.co.uk

Heat Line
(boilers and radiators)

Tel: 01773 596 611
E-mail: info@heatline.co.uk
Website: www.heatline.co.uk

Heat Spares Direct

Tel: 0844 357 0261
E-mail: shop@heatsparesdirect.co.uk
Website: www.heatsparesdirect.co.uk

Hepworth Heating
(see Glow-Worm)

Honeywell

Website: www.honeywell.com/uk

Hortsmann
(trading name of Secure Controls (UK) ltd)

Tel 0117 978 8700
Website: www.horstmann.co.uk

Hoval Ltd

Tel: 01636 672 711
E-mail: boilersales@hoval.co.uk
Website: www.hoval.co.uk

Ideal Heating

Tel: 01482 492 251
Website: www.idealheating.com

Invensys Controls in EMEA
(EMEA brands include Appliance EMEA,
Ranco, Drayton, Eberle & Eliwell)

Website: www.invensys.com

Johnson & Starley Ltd

Tel: 01604 762 881
Website: www.johnsonandstarleyltd.co.uk

Kamco ltd
(system treatment solutions)

Tel 01727 875 020
Website: www.kamco.co.uk

Kane International Ltd
(analysers)

Tel: 01707 375 550
Website: www.kane.co.uk

Keston Boilers

Tel: 01482 443 005
E-mail: info@keston.co.uk
Website: www.keston.co.uk

Kiwa GASTEC at CRE
(consultancy, training and product testing)

Tel 01242 677 877
E-mail: gastecenquiries@kiwa.co.uk
Website: www.kiwa.co.uk

Lanemark International Ltd
(burners)

Tel 024 7635 2000
Website: www.lanemark.com

Lennox Industries

Tel: 01604 669 100
Website: www.lennoxuk.com

Linde Industrial Gas
(Industrial gas supplier)

Website: www.linde-gas.com

Lister Gases Ltd
(LPG Supplier)

Tel: 0121 556 7181
E-mail: enquiries@listergases.co.uk
Website: www.listergases.co.uk

Lochinvar UK

Tel: 01295 269 981
Website: www.lochinvar.com

Magne-flo Excess Flowvalves ltd

Tel: 01564 782 776
Website: www.magne-flo.co.uk

Malvern Boilers Limited

Tel: 01684 893 777
E-mail: sales@malvernboilers.co.uk
Website: www.malvernboilers.co.uk

Man Diesel & Turbo
(compressors, turbines, etc.)

Tel: Germany +49 821 3220
E-mail: info-de@mandieselturbo.com

Mecserflex Manufacturing Company Ltd

Tel: 01793 773 320
E-mail: sales@mecserflex.co.uk
Website: www.mecserflex.co.uk

Midland Brass fittings Co ltd
(part of Midbras Group)

Tel: 0121 707 6666
E-mail: sales@midbras.co.uk
Website: www.midbras.co.uk

Monodraught ltd
(chimney specialists)

Tel: 01494 897 700
E-mail: info@monodraught.com
Website: www.monodraught.com

Murphy Pipelines
(civil engineering & utilities)

Tel: 020 7267 4366 (HQ)
E-mail: info@murphygroup.co.uk
Website: www.murphygroup.co.uk

Myson Radiators
(including heating controls)

Tel: 0845 402 3434
E-mail: sales@myson.co.uk
Website: www.myson.co.uk

Nordair Niche
(HVCA systems)

Tel: 0161 482 7900 (Northern office)
Tel: 01376 332 200 (Southern office)
E-mail: info@nordairniche.co.uk
Website: www.nordairniche.co.uk

Nu-Way ltd
(gas, oil & dual fuel burners)

Tel: 01905 794 331
E-mail: info@nu-way.co.uk
Website: wwwnu-way.co.uk

O.H. Ltd
(flue terminals)

Tel: 0800 644 4440 (technical)
E-mail: anthony@ohlimited.co.uk
Website: www.ohlimited.co.uk

Oso Hotwater (UK) Ltd
(manufacturer of stainless steel water
heaters)

Tel: 0191 482 0800
E-mail: technical@osohotwater.co.uk
Website: www.osohotwater.co.uk

Pactrol Controls ltd

Tel: 01942 529 240
Website: www.pactrol.com

Pegler Yorkshire
(manufacturer of plumbing & heating
products)

Tel: 0800 156 0050 (technical)
E-mail: tech.help@pegleryorkshire.co.uk
Website: www.pegler.co.uk

Polytank Ltd (one brand of Polygroup)

Tel: 01772 632 850
Website: www.polytank.co.uk

Polyethyiene Pipelines Company ltd
(utilities)

Tel: 0118 979 1596
E-mail: enquiries@polyethylenepipelines.co.uk
Website: www.polyethylenepipelines.co.uk

Potterton Myson Ltd

Tel: 0844 871 1560 (technical)
E-mail: info@potterton.co.uk
Website: www.potterton.co.uk

Powrmatic
(manufacturer of heating, cooling & air
condition systems)

Tel: 800 966 9100 (USA)
Website: www.powrmatic.com

Pumps & Motors (UK) Ltd
(Turney Turbines)

Tel: 020 8507 2288
E-mail: sales@pumpsmotors.co.uk
Website: www.pumpsmotors.co.uk

Ray Tube ltd
(part of Midtherm Flue Systems Ltd)

Tel: 01384 458 800
Website: midtherm.co.uk

Ravenheat Manufacturing Ltd

Tel: 0113 252 7007
Website: www.ravenheat.co.uk

Red bank (a brand of Hanson Heidelberg Cement Group)

Tel: 01530 270 333
Website: www.heidelbergcement.com

**Redland Roofing
(part of the Monier Group)**

Tel: 08708 702 595
Website: www.monier.co.uk

Regin Products Ltd

Tel: 01480 412 415
Website: www.reginproducts.co.uk

Remeha Commercial

Tel: 0118 978 3434
E-mail: technical@remeha.co.uk
Website: www.remeha.co.uk

Rettig Heating Ltd

Tel: 0191 492 1700
Website: www.purmoradson.com

Reznor UK Ltd

Tel: 01303 259 141
E-mail: info@reznor.co.uk
Website: www.reznor.co.uk

**Riello Ltd
(burners)**

Tel: 01480 432 144
E-mail: info@rielloburners.co.uk
Website: www.rielloburners.co.uk

**Roots Systems ltd
(boosters)**

Tel: 01453 826 581
E-mail: sales@roots-blowers.com
Website: www.roots-blowers.com

**Roberts Gordon UK Ltd
(radiant tube heaters)**

Tel: 0121 506 7700
E-mail: uksales@rg-inc.com
Website: www.rg-inc.com

**Safety Systems UK Ltd
(comprising of Bailey, Birkett, Amal, Marston & Marvac)**

Tel 0161 703 1999
E-mail: support@safteysystemsuk.com

Sangamo

Tel: 01475 745 131
E-mail: enquiries@sangamo.co.uk
Website: www.sangamo.co.uk

Schwank Ltd

Tel: 0208 641 3900
E-mail: sales@schwank.co.uk
Website: www.schwank.co.uk

**Selkirk (available via Deks Distribution UK)
(chimney systems)**

Tel: 01275 858 866
E-mail: sales@deks.org.uk
Website: www.selkirkchimney.co.uk

**Siemens Industry UK
(Landis and Staefa)**

Tel: 01276 696 000 (HQ)
Website: www.industry.siemens.co.uk

Siemens plc

Tel: 01276 696 000 (HQ)
Website: www.siemens.co.uk

Siemens Metering

Tel 0115 906 6000
Website: www.siemens.co.uk

SIT Controls UK

Tel: 0130 28 36 340
E-mail: situk@sitgroup.it
Website: www.sitgroup.it

Sperryn Gas Controls

Tel: 01744 611 811
E-mail: sperrynsales@cranebsu.com
Website: www.sperryn.co.uk

Space-Ray
(radiant heaters)

Tel 01473 830 551
E-mail: info@spaceray.co.uk
Website: www.spaceray.co.uk

Stadium
(ventilators)

Tel: 01843 854 002
E-mail: stadium@flambeau.com
Website: www.stadiumbuildingproducts.com

Stokvis Energy Systems

Tel: 0208 783 3050 (technical)
Website: www.stokvisboilers.com

Strebel Ltd

Tel: 01276 685 422
Website: www.strebel.co.uk

Sunvic Controls Ltd

Tel: 01698 810945 (technical)
E-mail: technical@sunvic.co.uk
Website: www.sunvic.co.uk

Teddington Control Co Ltd

Tel: 01726 222 505
Website: www.tedcon.com

Telegan
(test instruments & analysers)

Tel: 01235 557 700
E-mail: sales@telegan.co.uk
Website: www.telegangas.co.uk

Telford Copper Cylinders Ltd
(copper & stainless steel vessels)

Tel: 01952 257 961
E-mail: sales@telford-group.com
Website: www.telford-group.com

Testo Ltd
(test instruments & analysers)

Tel: 01420 544 433
Website: www.testo.co.uk

Test Products International Europe Ltd
(test instruments & analysers)

Tel: 01293 530 196
E-mail: contactus@tpieurope.com
Website: www.testproductsintl.com

Tower Flue Components
(part of TFC Group & includes Grasslin)

Tel: 01732 351 680
Website: www.tfc-group.co.uk

TracPipe (a brand of OmegaFlex Ltd)

Tel: 01295 676 670
E-mail: eurosales@omegaflex.net
Website: www.tracpipe.co.uk

Trianco

Tel 0114 257 2300
E-mail: info@trianco.co.uk
Website: www.trianco.co.uk

Vaillant Ltd

Tel: 0844 693 3133 (technical)
E-mail: technical@vaillant.co.uk
Website: www.vaillant.co.uk

Viessmann Ltd

Tel: 01952 675 070 (technical)
E-mail: technical-uk@viessman.com
Website: www.viessmann.com

Vokera Ltd

Tel: 0844 391 0999
E-mail: technical@vokera.co.uk
Website: www.vokera.co.uk

Wednesbury Copper Tube
(a brand of Mueller Industries Inc)
(copper tube manufacturer)

Tel: 01902 499 700
E-mail: sales@muellereurope.com
Website: www.wednesburytube.com

Wellman Group
(furnaces, kilns, boilers & process engineering)

Tel: 0121 543 0000
Website: www.wellman-group.com

White Rodgers (a brand of Emerson Climate Technologies)

Website: www.emersonclimate.com

Wilo (UK) Ltd

Website: www.wilo.co.uk

Worcester Bosch Group

Tel: 0844 892 3366 (technical)
E-mail: technical.enquiries@uk.bosch.com
Website: www.worcester-bosch.co.uk

Yorkshire Copper Tube ltd

Tel 0151 545 5107 (technical)
Website: www.yorkshirecopper.com

Zehnder Group UK Ltd
(commercial radiator & radiant heating)

Tel: 01252 515 151
Website: www.zehnder.co.uk

Licensed Gas Transporters

The companies listed below are licensed under section 7 of the Gas Act 1986 by Ofgem, to convey gas through pipes to premises in authorised areas specified in the licence or in any licence extensions.

British Gas Connections Ltd
30 The Causeway, Staines TW18 3BY

Tel: 01784 874 525
Fax: 01784 874 497

East Midlands Pipelines Ltd
Herald Way, Pegasus Business Park, East Midlands Airport, Castle Donnington DE74 2TU

Tel: 01332 393 327
Fax: 01332 393 027

ES Pipelines Ltd
Prospect Wells House, Outwood Lane, Chipstead, Coulsdon, Surrey CR5 3NA

Tel: 01737 558 378
Fax: 01737 558 315

GTC Pipelines Ltd
Woolpit Business Park, Woolpit, Bury St Edmunds, Suffolk IP30 9UQ

Tel: 01359 240 363
Fax: 01359 240 138

Independent Pipelines Ltd
Ocean Park House, East Tyndall Street, Cardiff CF24 5GT

Tel: 029 2030 4000
Fax: 029 2031 4140

Mowlem Energy Ltd
2 Redwood Court, Peel Park, East Kilbride, Glasgow G74 5PF

Tel: 01355 909 600
Fax: 01355 909 601

ScottishPower Gas Ltd

Commercial Operations, Power Systems,
New Alderston House, Dove Wynd, Strathclyde
Business Park, Bellshill ML4 3AD

Tel: 01698 413 295
Fax: 01698 413 064

SSE Pipelines Ltd

Westacott Way, Littlewick Green, Maidenhead,
Berks SL6 3QB

Tel: 01628 584 117
Fax: 01628 584 294

National Grid

Homer Road, Solihull, West Midlands B91 3LT

Tel: 0121 626 4431
Fax: 0121 623 2625

The Gas Transportation Company Ltd

Woolpit Business Park, Woolpit, Bury St
Edmunds, Suffolk IP30 9UQ

Tel: 01359 240 363
Fax: 01359 240 138

United Utilities Gas Networks

12th floor, Oakland House, Talbot Road,
Stretford, Manchester M16 0HQ

Tel: 0161 875 7042
Fax: 0161 875 7007

United Utilities Gas Pipelines Ltd

PO Box 3010, Links Business Park, Fortran
Road, St Mellons, Cardiff CF3 ODS

Tel: 029 2083 9250
Fax: 029 2083 9270

Utility Grid Installations Ltd

24a D'Olier Street, Dublin, Ireland

Tel: 00353 1 602 1061
Fax: 00353 1 602 1138

Organisations

Association for the Conservation of Energy (ACE)

Tel: 0207 359 8000
Fax: 0207 359 0863
Website: www.ukace.org

Association of Plumbing and Heating Contractors (APHC)

Tel: 0121 711 5030
Fax: 0121 705 7871
E-mail: info@aphc.co.uk
Website: www.competentpersonsscheme.co.uk

Boiler and Radiator Manufacturers Association Ltd (BARMA)

Tel: 0141 332 0826
Fax: 0141 332 5788

ICOM Energy Association (Formally BCEMA)

Tel: 01926 513 748
Fax: 01926 855 017
Website: www.icomenergyassociation.org.uk

British Marine Federation (BMF)

Tel: 01784 473 377
Fax: 01784 439 678
E-mail: info@britishmarine.co.uk
Website: www.britishmarine.co.uk

British Standards Institution (BSI)

Tel: 0208 996 9001
Fax: 0208 996 7001
E-mail: cservices@bsigroup.com
Website: www.bsigroup.com

Builders Merchants Federation (BMF)

Tel: 020 7439 1753
Fax: 020 7734 2766
E-mail: info@bmf.org.uk
Website: www.bmf.org.uk

Building Research Establishment (BRE)

Tel: 01923 664 000
Fax: enquiries@bre.co.uk
Website: www.bre.co.uk

Building Services Research & Information Association (BSRIA)

Tel: 01344 465 600
Fax: 01344 465 626
E-mail: bsria@bsria.co.uk
Website: www.bsria.co.uk

Chartered Institution of Building Services Engineers (CIBSE)

Tel: 020 8675 5211
Fax: 020 8675 5449
Website: www.cibse.org

Chartered Institute of Plumbing and Heating Engineering (CIPHE)

Tel: 01708 472 791
Fax: 01708 448 987
E-mail: info@ciphe.org.uk
Website: www.ciphe.org.uk

City and Guilds of London Institute

Tel: 0844 543 0033
Website: www.cityandguilds.com

Combined Heat and Power Association (CHPA)

Tel: 020 7828 4077
Fax: 020 7828 0310
E-mail: info@chpa.co.uk
Website: www.chpa.co.uk

Confederation of British Industrys (CBI)

Tel: 020 7379 7400
Fax: 020 7379 7200
Website: www.cbi.org.uk

Construction Skills

Website: www.cskills.org

Consumer Direct (handles energy issues that may be aimed at Ofgem)

Tel: 08454 04 05 06

Consumer Focus
(formed by the mergers of English, Scottish and Welsh National Consumer Councils and Postwatch and Energywatch)

Tel: 020 7799 7900
Fax: 020 7799 7901
Website: www.consumerfocus.org.uk

Department for Business Innovation & Skills (formerly BERR, and before that, the DTI)

Tel: 020 7215 5000, or
Tel: 020 7215 6740
Website: www.bis.gov.uk

Energy & Utility Skills

Tel: 0845 077 99 22
Fax: 0845 077 99 33
E-mail: enquiries@euskills.co.uk
Website: www.euskills.co.uk

Energywatch (see Consumer Focus)

Energy Savings Trust (EST)

Tel: 020 7222 0101
Website: www.energysavingtrust.org.uk

Federation of Master Builders (FMB)

Tel: 020 7242 7583
Fax: 020 7404 0296
Website: www.fmb.org.uk

Gas Industry Safety Group (GISG)

Tel: 020 7706 5108
E-mail: info@gisg.org.uk
Website: www.gisg.org.uk

Gas Safe Register®

Tel: 0800 408 5500
E-mail: enquiries@gassaferegister.co.uk
Website: www.gassaferegister.co.uk

HSE (Health and Safety Executive) Books

Tel: 01787 881 165
Fax: 01787 313 995
E-mail: hsebooks@prolog.uk.com
Website: www.books.hse.gov.uk

Heating & Hotwater Industry Council (HHIC)

Website: www.centralheating.co.uk

HVCA - See The Building & Engineering Services Association

ICOM Energy Association (Formally BCEMA)

Tel: 01926 513 748
Fax: 01926 855 017
Website: www.icomenergyassociation.org.uk

Institute of Domestic Heating & Environmental Engineers

Tel: 02380 66 89 00
Fax: 02380 66 08 88
E-mail: admin@idhee.org.uk
Website: www.idhee.org.uk

Institution of Gas Engineers and Managers (IGEM)

Tel: 0844 375 4436
Fax: 01509 678 198
E-mail: general@igem.org.uk
Website: www.igem.org.uk

National Association of Chimney Sweeps (NACS)

Tel: 01785 811 732
Fax: 01785 811 712
Website: www.nacs.org.uk

National Caravan Council (NCC)

Tel: 01252 318 251
Fax: 01252 322 596
E-mail: info@thencc.org.uk
Website: www.nationalcaravan.co.uk

National Federation of Builders (NFB)

Tel: 08450 578 160
Fax: 08450 578 161
Website: www.builders.org.uk

National Housing Federation

Tel: 020 7067 1010
Fax: 020 7067 1011
Website: www.housing.org.uk

National Inspection Council for Electrical Installations Contracting (NICEIC)

Tel: 0870 013 0391 (technical)
Website: www.niceic.com

Ofgem (Office of the Gas and Electricity Markets) – see also Consumer Direct

Tel: 020 7901 7295
Fax: 020 7901 7066
Website: www.ofgem.gov.uk

Ofwat (the Water Services Regulation Authority)

Tel: 0121 644 7500
Fax: 0121 644 7559
E-mail: mailbox@ofwat.gsi.gov.uk
Website: www.ofwat.gov.uk

Royal Society for the Prevention of Accidents (RoSPA)

Tel: 0121 248 2000
Website: www.rospa.com

Scottish and N. I. Plumbing Employers' Federation (SNIPEF)

Tel: 0131 225 2255
Fax: 0131 226 7638
E-mail: info@snipef.org
Website: www.snipef.org.uk

Society of British Gas Industries (sbgi) – has two divisions; HHIC (see HHIC in this Part) and sbgi Utility Networks

Tel: 01926 513 777
Fax: 01926 511 923
E-mail: mail@sbgi.org.uk
Website: www.sbgi.org.uk

sbgi Utility Networks

Tel: 01926 513 765
Fax: 01926 857 474
Website: www.sbgi.org.uk/utillitynetworks

Stove and Fireplace Advice (formerly the National Fireplace Association)

Tel: 01494 411 242 (Ext 2), or
Tel: 0845 643 1901
Fax: 0870 130 6747
E-mail: advice@stoveandfireplaceadvice.org.uk
Website: www.stoveandfireplaceadvice.org.uk

The British Electrotechnical and Allied Manufacturers Association

Tel: 0207 793 3011 (technical)
E-mail: technical@beama.org.uk
Website: www.beama.org.uk

The Building & Engineering Services Association (B&ES) – formerly the HVCA

Tel: 020 7313 4900
Fax: 020 7727 9268
E-mail: contact@hvca.org.uk
Website: www.hvca.org.uk

The Energy Institute

Tel: 0207 467 7100
E-mail: info@energyinst.org
Website: www.energyinst.org

The Stationary Office (TSO)

Tel: 0870 600 5522
Website: www.tso.co.uk

United Kingdom Accreditation Service (UKAS)

Tel: 020 8917 8400
E-mail: info@ukas.com
Website: www.ukas.com

UKlpg (merger between LP Gas Association and the Association for Liquid Gas Equipment and Distributors)

E-mail: mail@uklpg.org
Website: www.uklpg.org

Nationally Accredited Certification Scheme (ACS) for Individual Gas Fitting Operatives

UKAS ACCREDITED CERTIFICATION BODIES

Blue Flame Certification
Unit 13 & 14, Chatterley Whitfield Enterprise Centre, Chatterley Whitfield, Stoke on Trent, Staffordshire, ST6 8UW

Tel: 0845 194 90 31
Website: www.blueflamecertification.com

BPEC
2 Mallard Way, Pride Park, Derby, DE24 8GX

Tel: 0845 644 6558
Fax: 0845 121 1931
E-mail: info@bpec.org.uk
Website: www.bpec.org.uk

Construction Skills
Building Engineering Services (BES),
Bircham Newton, Kings Lynn, Norfolk,
PE31 6RH

Tel:	0300 456 7700
Tel:	0344 994 4133 (certification scheme)
E-mail:	bes.enquiry@cskills.org
Website:	www.cskills.org

LOGIC Certification Limited
Unit 2, 1 Rowdell Road, Northolt, Middlesex,
UB5 5QR

Tel:	020 8839 2439
E-mail:	enquiries@logic-cert.com
Website:	www.logiccertification.com

NIC Certification (formerly Zurich Certification)
Whitting Valley Road, Old Whittington,
Chesterfield, S41 9EY

Tel:	0500 600 545
Tel:	01246 269 048 (main office)
Fax:	01246 269 049
E-mail:	certification@niccertification.com
Website:	www.niccertification.com

UK Certification
Unit 5, Station Close, Westover Trading Estate,
Langport, Somerset, TA10 9RB

Tel:	01458 252 757
E-mail:	info@ukcertification.co.uk
Website:	www.ukcertification.org

Health and Safety Executive (HSE) area offices

London
Rose Court

2 Southwark Bridge, London, SE1 9HS.
Fax: 020 7556 2102

Covers: London only.

Westminster Office

Caxton House, Tothill Street, London, SW1H 9NA.
Fax: 020 7227 3802

Wales & South West
Cardiff

Government Buildings, Phase 1, Ty Glas,
Llanishen, Cardiff CF14 5SH.
Fax: 029 2026 3120

Covers: Merthyr Tydfil, Rhondda Cynon Taff, Vale
of Glamorgan, Bridgend, Neath Port Talbot,
Powys, Blaenau Gwent, Caerphilly, Cardiff,
Monmouthshire, Newport, Torfaen and part of
Powys.

Bristol

4th Floor, The Pithay, All Saints Street, Bristol,
BS1 2ND.
Fax: 01179 262 998

Covers: Bristol, Somerset, Bath and North East
Somerset, North Somerset, Gloucestershire, South
Gloucestershire, Dorset, Swindon and Wiltshire.

Plymouth

North Quay House, Sutton Harbour, Plymouth,
PL4 0RA.
Fax: 01752 226 024

Covers: Devon and Cornwall.

Clwyd

Unit 7 & 8 Edison Court, Ellice Way, Wrexham Technology Park, Wrexham, Clwyd, LL13 7YT.
Fax: 01978 355 669

Covers: Anglesey, Conwy, Denbighshire, Flintshire, Wrexham, Gwynedd and part of Powys.

Dyfed

Tŷ Myrddin, Old Station Road, Carmarthen, Carmarthenshire, SA31 1LP.
Fax: 01267 223 267

Covers: Carmarthenshire, Pembrokshire, Ceredigion and Swansea.

Dorset

14 New Fields, Stinsford Road, Nuffield Industrial Estate, Poole, Dorset, BH17 0NF.
Fax: 01202 667 224

Covers: Dorset

East & South East

Basingstoke

Priestley House, Priestley Road, Basingstoke, RG24 9NW.
Fax: 01256 404 100

Covers: Berkshire, Hampshire, Oxfordshire and Isle of Wight.

Bedford

Woodlands, Manton Lane, Manton Lane Industrial Estate, Bedford.
Fax: 01234 220 633

Covers: Hertfordshire, Cambridgeshire, Bedfordshire and Milton Keynes and Buckinghamshire.

Chelmsford

Wren House, Hedgerows Business Park, Colchester Road, Springfield, Chelmsford, CM2 5PF.
Fax: 01245 706 222

Covers: Essex (except Barking, Havering, Redbridge and Waltham Forest, these are covered by London), Norfolk and Suffolk.

East Grindstead

Phoenix House, 23-25 Cantelupe Road, East Grinstead, West Sussex, RH19 3BE.
Fax: 01342 334 222

Covers: East & West Sussex and Surrey.

Norwich

Rosebery Court, 2nd Floor, St Andrew's Business Park, Norwich, Norfolk, NR7 0HS.

Covers: Norfolk, Suffolk and Essex (see also Chelmsford).

Kent

International House, Dover Place, Ashford, Kent, TN23 1HU.
Fax: 01233 634 827

Covers: Kent.

Midlands

Birmingham

1 Hagley Road, Birmingham, B16 8HS.
Fax: 0121 607 6349.

Covers: West Midlands.

Northampton

900 Pavilion Drive, Northampton Business Park, Northampton, NN4 7RG.
Fax: 01604 738333.

Covers: Leicestershire, Northampton, Rutland and Warwickshire.

Nottingham

City gate West, Level 6 (First Floor), Toll House Hill, Nottingham, NG1 5AT.
Fax: 0115 971 2802

Covers: Nottinghamshire, Derbyshire, Lincolnshire (North Lincolnshire covered by Sheffield Office).

Stoke on Trent

Lyme Vale Court, Lyme Drive, Parklands Business Park, Newcastle Road, Trent Vale, Stoke on Trent, ST4 6NW.
Fax: 01782 602 400

Covers: Staffordshire and Shropshire.

Worcester

Haswell House, St Nicholas Street, Worcester, WR1 1UW.
Fax: 01905 723 045

Covers: Worcestershire and Herefordshire.

North West

Bootle Headquarters

Redgrave Court, Merton Road, Bootle, Merseyside, L20 7HS.

Manchester

Grove House, Skerton Road, Manchester, M16 0RB.
Fax: 0161 952 8222

Covers: Merseyside, Cheshire and Greater Manchester.

Carlisle

2 Victoria Place, Carlisle, CA1 1ER.
Fax: 01228 548 482

Covers: Cumbria.

Yorkshire and North East

Leeds

The Lateral, 8 City Walk, Leeds, LS11 9AT.
Fax: 0113 283 4382.

Covers: West and North Yorkshire.

Sheffield

Health and Safety Executive, Foundry House, 3 Millsands, Riverside Exchange, Sheffield, S3 8NH.
Fax: 0114 291 2379.

Covers: East Yorkshire, North Lincolnshire, NE Lincolnshire, South Yorkshire and Kingston-upon-Hull.

Newcastle

Alnwick House, Benton Park View, Newcastle-Upon-Tyne, NE98 1YX.
Fax: 0191 202 6300

Covers: Northumberland, Tyne and Wear, Durham and Cleveland.

Scotland

Edinburgh

Belford House, 59 Belford Road, Edinburgh, EH4 3UE.
Fax: 0131 247 2121

Covers: Borders, Lothian, Central Perth, Kinross, Fife and Dundee.

Glasgow

1st Floor, Mercantile Chambers,
53 Bothwell Street, Glasgow, G2 6TS.
Fax: 0141 275 3100

Covers: West Scotland.

Aberdeen

Field Operations Division, Lord Cullen House,
Fraser Place, Aberdeen, AB25 3UB.
Fax: 01224 252 525

Covers: Angus, Aberdeenshire, Moray and Shetland.

Inverness

Longman House, 28 Longman Road,
Longman Industrial Estate, Inverness, IV1 1SF.
Fax: 01463 713 459.

Covers: Highlands and Orkney.

Statutory and Normative documents relevant to gas work

Statutory Regulations

*   Health and Safety at Work etc Act 1974
*   The Management of Health and Safety at Work Regulations 1999
*   The Building Regulations (England & Wales) 2010
*   The Building (Scotland) Amendment Regulations 2011
*   The Building (Amendment No. 2) Regulations (Northern Ireland) 2010
*   Building Control (Approved Documents) Order 2007 (as applied to the Isle of Man)
*   The Gas Safety (Installation & Use) Regulations 1998
*   Gas Safety (Installation & Use) Regulations (Northern Ireland) 2004
*   Gas Safety (Installation & Use) Regulations 1994 as amended and applied by the Gas Safety (Application) Order 1996 (Isle of Man)
*   The Gas Safety (Management) Regulations 1996

Available from TSO (see contacts list).

Health and Safety Commission

*   Standards of Training in Safe Gas Installation – Approved Code of Practice
*   Gas Safety (Installation and Use) Regulations – Approved Code of Practice and Guidance
*   CS4 The Keeping of LPG in Cylinders and Similar Containers
*   CS11 The Storage and Use of LPG at metered Estates

Available from HSE Books (see Organisations list).

British Standards

BS 10

Specification for flanges and bolting for pipes, valves, and fittings

BS 21

Specification for pipe threads for tubes and fittings where pressure-tight joints are made on threads (metric dimensions)

Note: BS 21 has been partially replaced by BS EN 10226-1: Pipe threads where pressure tight joints are made on the threads. Taper external threads and parallel internal threads. Dimensions, tolerances and designation

BS 143 & 1256

Threaded pipe fittings in malleable cast iron and cast copper alloy. – Comparable standard BS EN 10242: Threaded pipe fittings in malleable cast iron

BS 669-2

Flexible hoses, end fittings and sockets for gas burning appliances. Specification for corrugated metallic flexible hoses, covers, end fittings and sockets for catering appliances burning 1st, 2nd and 3rd family gases

BS 1560

Circular flanges for pipes, valves and fittings (class designated)

Note: BS 1560, parts 3.1 & 3.3 have been replaced by BS EN 1759-1: Flanges and their joints. Circular flanges for pipes, valves, fittings and accessories, class-designated. Steel flanges, NPS 1/2 to 24

BS 1832

Specification for compressed asbestos fibre jointing

Note: BS 1832 has been withdrawn.

BS 2051

Tube and pipe fittings for engineering purposes. Copper and copper alloys capillary and compression tube fittings for engineering purposes

Note: BS 2051 has been declared as being obsolescent.

BS 2971

Specification for Class II arc welding of carbon steel pipework for carrying fluids

BS 3212

Specification for flexible rubber tubing, rubber hose and rubber hose assemblies for use in LPG vapour phase and LPG/air installations

BS 3974-1

Specification for pipe supports. Pipe hangers, slider and roller type supports

Note: BS 3974-1 has been withdrawn.

BS 3974-2

Specification for pipe supports. Pipe clamps, cages, cantilevers and attachments to beams

Note: BS 3974-2 has been withdrawn.

BS 4089

Specification for metallic hose assemblies for liquid petroleum gases and liquefied natural gases

BS 4368-1

Metallic tube connectors for fluid power and general use. Split collet compression fittings

BS 5114

Specification for performance requirements for joints and compression fittings for use with polyethylene pipes

Note: BS 5114 has been declared as being obsolescent.

BS 5292

Specification for jointing materials and compounds for installations using water, low-pressure steam or 1st, 2nd and 3rd family gases.

Note: BS 5292 has been partially replaced by BS 6956-1, 6956-5, 6956-6 and 6956-7

BS 5440

Flueing and ventilation for gas appliances of rated input not exceeding 70kW net (1st, 2nd and 3rd family gases):

Part 1:   Specification for the installation of gas appliances to chimneys and for maintenance of chimneys

Part 2:   Specification for the installation and maintenance of ventilation for gas appliances

BS 5449

Specification for forced circulation of hot water central heating systems for domestic premises.

Note: BS 5449 has been partially replaced by BS EN 12828: Heating systems in buildings. Design for water-based heating systems, BS EN 12831: Heating systems in buildings. Method for calculation of the design heat load and BS EN 14336: Heating systems in buildings. Installation and commissioning of water based heating systems.

BS 5482-1

Code of practice for domestic butane- and propane-gas-burning installations. Installations at permanent dwellings, residential park homes and commercial premises, with installation pipework sizes not exceeding DN 25 for steel and DN 28 for corrugated stainless steel or copper

BS 5482-2

Domestic butane and propane – gas burning installations. Installations in caravans and non-permanent dwellings.

Note: BS 5482: Part 2 has been partially replaced by BS 5482-1 and BS EN 1949: Specification for the installation of LPG systems for habitation purposes in leisure accommodation vehicles and accommodation purposes in other vehicles.

PD 5482-3

Code of practice for domestic butane- and propane-gas-burning installations. Installations in boats, yachts and other vessels

Note: PD 5482-3 replaces the previous BS 5482-3.

BS 5546

Specification for installation and maintenance of gas-fired water-heating appliances of rated input not exceeding 70kW (net).

BS 5854

Code of practice for flues and flue structures in buildings

BS 5864

Installation and maintenance of gas-fired ducted air heaters of rated input not exceeding 70 kW net (2nd and 3rd family gases). Specification.

BS 5885-1

Automatic gas burners. Specification for burners with input rating 60kW and above

Note: BS 5885 has been partially replaced by BS EN 676: Automatic forced draught burners for gaseous fuels.

BS 5978

Safety and performance of gas-fired hot water boilers (60kW to 2MW input). Specification for general requirements

Note: BS 5978 has been partially replaced by BS EN 656: Gas-fired central heating boilers. Type B boilers of nominal heat input exceeding 70 kW but not exceeding 300 kW

BS 6004

Electric cables. PVC insulated, non-armoured cables for voltages up to and including 450/750V, for electric power, lighting and internal wiring

BS 6007

Electric cables. Single core unsheathed heat resisting cables for voltages up to and including 450/750 V, for internal wiring

BS 6173

Specification for installation of gas-fired catering appliances for use in all types of catering establishments (2nd and 3rd family gases)

BS 6230

Specification for installation of gas-fired forced convection air heaters for commercial and industrial space heating (2nd and 3rd family gases)

BS 6231

Electric cables. Single core PVC insulated flexible cables of rated voltage 600/1000 V for switchgear and controlgear wiring.

BS 6400

Specification for installation, exchange, relocation and removal of gas meters with a maximum capacity not exceeding 6 m³/h –

Part 1:   Low pressure (2nd family gases)
Part 2:   Medium pressure (2nd family gases)
Part 3:   Low and medium pressure
           (3rd family gases)

BS 6501-1

Metal hose assemblies. Guidance on the construction and use of corrugated hose assemblies

BS 6501-2

Flexible metallic hose assemblies. Specification for strip wound hoses and hose assemblies

Note: BS 6501-2 has been superseded by BS EN ISO 15465: Pipework. Stripwound metal hoses and hose assemblies.

BS 6644

Specification for the Installation and maintenance of gas-fired hot water boilers of rated inputs between 70kW (net) and 1.8 MW (net) (2nd and 3rd family gases)

BS 6798

Specification for installation and maintenance of gas-fired boilers of rated input not exceeding 70kW net.

BS 6891

Installation of low pressure gas pipework of up to 35 mm (R11/4) in domestic premises (2nd family gas). Specification.

BS 6896

Specification for installation and maintenance of gas-fired overhead radiant heaters for industrial and commercial heating (2nd and 3rd family gases)

BS 6956

Jointing materials and compounds. Specification for corrugated metal joint rings

BS 7244

Specification for flame arresters for general use

Note: BS 7244 is current but proposed for withdrawal.

BS 7461

Specification for electrically operated automatic gas shut-off valves fitted with throughput adjusters, proof of closure switches, closed position indicator switches or gas flow control

BS 7531

Rubber bonded fibre jointing for industrial and aerospace purposes. Specification.

BS 7671

Requirements for electrical installations. IET Wiring Regulations. Seventeenth Edition

BS 8313

Code of practice for accommodation of building services in ducts

BS 9999

Fire safety code of practice for the design, management and use of buildings.

BS EN 161

Automatic shut-off valves for gas burners and gas appliances

BS EN 656

Gas-fired central heating boilers. Type B boilers of nominal heat input exceeding 70kW, but not exceeding 300kW

BS EN 676

Automatic forced draught burners for gaseous fuels.

BS EN 751

Sealing materials for metallic threaded joints in contact with 1st, 2nd and 3rd family gases and hot water.

Part 1:   Anaerobic jointing compounds.
Part 2:   Non-hardening jointing compounds
Part 3:   Unsintered PTFE tape

BS EN 1254-1

Copper and copper alloys. Plumbing fittings. Fittings with ends for capillary soldering or capillary brazing to copper tubes.

BS EN 1254-2

Copper and copper alloys. Plumbing fittings. Fittings with compression ends for use with copper tubes

BS EN 1443

Chimneys. General requirements

BS EN 1514

Flanges and their joints. Dimensions of gaskets for PN-designated flanges

BS EN 1555

Plastics piping systems for the supply of gaseous fuels. Polyethylene (PE). General

Note: BS EN 1555-1 replaces the previous BS 7281: Specification for polyethylene pipes for the supply of gaseous fuels.

BS EN 1555-3

Plastics piping systems for the supply of gaseous fuels. Polyethylene (PE). Fittings.

Note: BS EN 1555-3 replaces the previous BS 7336: Specification for polyethylene fusion fittings with integral heating elements for use with polyethylene pipes for the conveyance of gaseous fuels

BS EN 1092-3

Flanges and their joints. Circular flanges for pipes, valves, fittings and accessories, PN designated. Copper alloy flanges

BS EN 1993-3-2

Eurocode 3. Design of steel structures. Towers, masts and chimneys. Chimneys

Note: BS EN 1993-3-2 replace BS 4076: Specification for steel chimneys.

BS EN 10048

Hot rolled narrow steel strip. Tolerances on dimensions and shape

BS EN 10051

Continuously hot-rolled strip and plate/sheet cut from wide strip of non-alloy and alloy steels. Tolerances on dimensions and shape.

BS EN 10095

Heat resisting steels and nickel alloys.

BS EN 10241

Steel threaded pipe fittings.

BS EN 10253-1

Butt-welding pipe fittings. Wrought carbon steel for general use and without specific requirements

BS EN 10226-1

Pipe threads where pressure tight joins are made on the threads. Taper external threads and parallel internal threads. Dimensions, tolerances and designation.

BS EN 10242

Threaded pipe fittings in malleable cast iron.

BS EN 12864

Low-pressure, non adjustable regulators having a maximum outlet pressure of less than or equal to 200mbar, with a capacity of less than or equal to 4kg/h, and their associated safety devices for butane, propane or their mixtures

BS EN 13410

Gas-fired overhead radiant heaters. Ventilation requirements for non-domestic premises.

BS EN 1555

Plastics piping systems for the supply of gaseous fuels. Polyethylene (PE) –

Part 1:   General
Part 2:   Pipes
Part 3:   Fittings

BS EN 50379

Specification for portable electrical apparatus designed to measure combustion flue gas parameters of heating appliances –

Part 1:   General requirement and test methods

Part 2:   Performance requirements for apparatus used in statutory inspections and assessment

Part 3:   Performance requirements for apparatus used in non-statutory servicing of gas fired heating appliances

Note: BS EN 50379 replaces the previous BS 7927: Heating appliances for domestic applications. Portable apparatus designed to detect and measure specific combustion flue gas products.

BS EN 60079-1

Explosive atmospheres. Equipment protection by flameproof enclosures "d".

Note: BS EN 60079-1 replaces the previous BS EN 50018: Electrical apparatus for potentially explosive atmospheres. Flameproof enclosure 'd'.

BS EN 62305-1

Protection against lightning. General principles.

Note: Replaces BS 6651: Code of practice for protection of structures against lightning.

BS EN ISO 9445-1

Continuously cold-rolled stainless steel. Tolerances in dimensions and form. Narrow strip and cut lengths.

Note: BS EN ISO 9445-1 replaces the previous BS EN 10258: Cold-rolled stainless steel narrow strip and cut lengths. Tolerances on dimensions and shape.

BS EN ISO 10380

Pipework. Corrugated metal hoses and hose assemblies

BS EN ISO 15465

Pipework – Stripwound metal hoses and hose assemblies

BS EN ISO 18286

Hot-rolled stainless steel plates. Tolerances on dimensions and shape

Note: BS EN ISO 18286 replaces the previous BS EN 10029: Specification for tolerances on dimensions, shape and mass for hot rolled steel plates 3mm thick or above

Available from the British Standards Institute (see **Organisations** in this Part).

Uklpg

Codes of Practice

COP1   Bulk LPG storage at fixed installations.

Part 1: Design, installation and operation of vessels located above ground

Part 2: Small bulk installations for domestic purposes

Part 3: Examination and inspection

Part 4: Buried/mounded LPG storage vessels

COP4   Safe and satisfactory operation of propane-fired thermoplastic and bitumen boilers, mastic asphalt cauldrons/mixer, hand tools and similar equipment.

COP9   LPG-air plants.

COP10   Containers attached to mobile gas-fired equipment.

COP11   Autogas installation.

COP12   Recommendations for safe practice in the design and operation of LPG cylinder filling plants.

COP17   Purging LPG vessels and systems.

COP18   Safe use of LPG as a propulsion fuel for boats, yachts and other craft.

COP21   Guidelines for caravan ventilation and flueing checks.

COP22   LPG Piping system design and installation.

COP24   Use of LPG cylinders –

Part 1: Use of LPG cylinders at residential and similar premises

Part 3: The use of LPG in mobile catering vehicles and similar commercial vehicles

Part 4: The use of LPG for catering and outdoor functions

Part 5: The storage and use of LPG on construction sites

Part 6: The use of propane in cylinders at commercial and industrial premises

COP25   LPG Central storage and distribution systems for multiple consumers.

COP30   Gas installations for motive power on mechanical handling and maintenance equipment

Note: The list is not exhaustive

Available from UKLPG (see **Organisations** list).

Institution of Gas Engineers and Managers (IGEM) – Utilization Procedures

IGE/UP/1 (Edition 2) Strength testing, tightness testing and direct purging of industrial and commercial gas installations

IGE/UP/1A (Edition 2) Strength testing, tightness testing and direct purging of small, low-pressure industrial and commercial natural gas installations

IGE/UP/1B (Edition 3) Tightness testing and direct purging of small Liquefied Petroleum Gas/Air, Natural Gas and Liquefied Petroleum Gas installations

IGEM/UP/2 (Edition 2) Installation pipework on industrial and commercial premises

IGE/UP/3 (Edition 2) Gas fuelled spark ignition and dual fuel engines

IGEM/UP/4 (Edition 3) Commissioning of gas fired plant on industrial and commercial premises

IGEM/UP/6 (Edition 2) Application of compressors to natural gas fuel systems

IGE/UP/7 (Edition 2) Gas Installations in timber framed and light steel framed buildings

IGE/UP/9 (Edition 2) Application of Natural Gas and fuel oil systems to gas turbines and supplementary and auxiliary fired burners

IGE/UP/10 (Edition 3 Including amendments October 2010) Installation of flued gas appliances in industrial and commercial premises

IGEM/UP/11 (Edition 2) Gas installations in educational establishments

IGE/UP/12 Application of burners and controls to gas fired process plant

IGEM/UP/17 Shared chimney and flue systems for domestic gas appliances

IGEM/UP/18 Gas installations for vehicle repair and body shops

Note: The list is not exhaustive

Available from the IGEM
(see **Organisations** list)

CORGI*direct* Publications – 13

Gas – Domestic

Manual Series

GID1 Essential Gas Safety
(Fifth Edition – Second Revised)

GID2 Gas Cookers and Ranges
(Third Edition)

GID3 Gas Fires and Space Heaters
(Fourth Edition)

GID4 Laundry, Leisure and Refrigerators
(out of print)

GID5 Water Heaters
(Second Edition)

GID6 Gas Meters
(Third Edition)

GID7 Central Heating – Wet and Dry
(Fourth Edition)

GID8 Gas Installations in Timber/Light Steel
Frame Buildings
(Second Edition – Second Revised)

GID9 LPG – Including Permanent Dwellings,
Leisure Accommodation Vehicles,
Residential Park Homes and Boats
(Third Edition – Second Revised)

GID11 Using Portable Electronic Combustion
Gas Analysers for Investigating
Reports of Fumes (First Edition)

GID12 Using Portable Electronic Combustion
Gas analyser – Servicing and
Maintenance (First Edition)

FFG2 Fault Finding – wet central heating
systems Domestic (First Edition)

Pocket Series

USP1 The Gas Industry Unsafe Situations
Procedure (Sixth Edition)

CPA1 Combustion performance testing –
Domestic (First Edition)

SRB1 Ventilation Slide Rule
(Third Edition – Third Revised)

GRB1 Gas Rating Slide Rule Natural Gas –
Domestic (Second Edition – Revised)

GRB2 Gas Rating Slide Rule LPG (Propane) –
Domestic (First Edition)

TTP1 Tightness Testing and Purging
(Second Edition – Second Revised)

FFG1 Fault Finding Guide (out of print)

TTG1 Terminals and Terminations
(Fourth Edition – Revised)

Design Guide

WAH1 Warm Air Heating System
Design Guide (out of print)

Forms

All CORGI*direct's* gas forms carry Gas Safe Register® logo under licence from the HSE.

CP1	Gas Safety Record
CP2	Leisure Industry Landlord's Gas Safety Record
CP3FORM	Chimney/Flue/Fireplace and Hearth Commissioning Record
CP4	Gas Safety Inspection
CP6	Service/Maintenance Checklist
CP9	Visual Risk Assessment of Gas Appliances
CP12	Landlord/Home Owner Gas Safety Record
CP14	Warning/Advice Notice
CP26	Fumes Investigation Report
CP32	Gas Testing and Purging – Domestic (NG)
CP43	Risk Assessment for Existing Chimney Systems in Voids Where Inadequate Access for Inspection is Provided

Labels

CP3PLATE	Chimney/Hearth Notice Plate
WLID	Immediately Dangerous Warning labels/tags
WLAR	At Risk Warning labels/tags
TG5	Tie on Uncommissioned Appliance labels
TG8	Void Property Tag
WL5	Gas Emergency Control Valve labels
WL8	Compartment/Ventilation labels
WL9	Electrical Bonding labels
WL13	Serviced By Label

Gas – Non-Domestic

ND1	Essential Gas Safety Non-domestic (Second Edition – Second Revised)
ND2	Commercial Catering and Laundry Non-domestic (Second Edition)
ND3	Commercial Heating Non-domestic (First Edition)

Pocket Guides

USP1	The Gas Industry Unsafe Situations Procedure (Sixth Edition)
CPA2	Combustion performance testing – Non Domestic (First Edition)
VENT1	Boiler Ventilation – Non Domestic (First Edition)

Forms

CP15	Plant Commissioning/Servicing Record (Non-domestic)
CP16	Gas Testing and Purging (Non-domestic)
CP17	Gas Installation Safety Report (Non-domestic)
CP42	Gas Safety Inspection (Commercial Catering Appliances)
CP44	Mobile Catering Vehicle/Trailer Safety Check

Labels

WLID	Immediately Dangerous Warning labels/tags
WLAR	At Risk Warning labels/tags
WL10	Emergency Control Valve labels
WL35	Manual Gas Isolation Valve
WL36	Automatic Gas Isolation Valve

Electrical

Form

CP22   Minor Electrical Installation
Works Certificate

Plumbing

Manual Series

HEM1   Hygiene Engineering (First Edition)

Pocket Guides

CDP1   Commissioning of Water Pipework –
Domestic (First Edition – Revised)

CWCB   Cleansing of Wet central heating
Systems (First Edition)

Design Guides

WCH1   Wet Central Heating System
Design Guide (out of print)

UVDG   Unvented Hot Water Systems
Design Guide (out of print)

Forms

CP8   Domestic Unvented Hot Water
Storage Vessel
Commissioning/Inspection Record

CP20   Central Heating
Commissioning/Inspection Record

CP33   Commissioning of Water Pipework

CP34   Central Heating Cleansing Record

CP40   Bathroom Quality Check Sheet

CP41   Combined Pressure Test Record Sheet

Label

TG9   Cleansing service label

Renewables

Manual Series

EEM1   Ground Source Heat Pumps
(First Edition)

EEM2   Domestic Solar Hot Water Systems
(First Edition)

EEM3   Domestic Biomass Systems
(First Edition)

Form

CP11   Solar Thermal Commissioning Record

Labels

WL33   Solar Thermal Installation

WL34   Solar Thermal Pressure Relief Valve

Business

Forms

CP10   Contract of Work & Notice of the
Right to Cancel

CP19   Invoice form

Notice

NA1   Sorry We Missed You Cards